ELECTROMAGNETIC THEORY
PROBLEMS AND SOLUTIONS

ELECTROMAGNETIC THEORY

PROBLEMS AND SOLUTIONS

Volume 2

K. FOSTER

Department of Electrical Engineering, Lanchester College of Technology, Coventry

R. ANDERSON

Department of Electrical Engineering, Lanchester College of Technology, Coventry

St. Martin's Press

ENGLAND:	BUTTERWORTH & CO. (PUBLISHERS) LTD. LONDON: 88 Kingsway, WC2B 6AB
AUSTRALIA:	BUTTERWORTH & CO. (AUSTRALIA) LTD. SYDNEY: 20 Loftus Street MELBOURNE: 343 Little Collins Street BRISBANE: 240 Queen Street
CANADA:	BUTTERWORTH & CO. (CANADA) LTD. TORONTO: 14 Curity Avenue, 374
NEW ZEALAND:	BUTTERWORTH & CO. (NEW ZEALAND) LTD. WELLINGTON: 49/51 Ballance Street AUCKLAND: 35 High Street
SOUTH AFRICA:	BUTTERWORTH & CO. (SOUTH AFRICA) (Pty) LTD. DURBAN: 33/35 Beach Grove

Library of Congress Catalog Card Number 76–123637

Made and printed in Great Britain

PREFACE

The production of this second volume of problems stems from the same reasons which led to the first volume. The authors believe that this subject, like others, cannot be thoroughly understood by a student unless he spends time solving problems. Unfortunately for the student, the ability to solve problems comes with understanding: to some extent a vicious circle. The object of this collection of problems and solutions is to enable the student to break out of this circle. It is not intended as a teaching text, although something can be learnt from the study of solutions. Primarily, the student should rely on both his lecture notes and a wide reading of some of the excellent texts available; he should then be in a position to attempt some of the problems in this book. At the outset he will probably find it difficult to start the problem and the authors would suggest that (after having spent some time with the problem) he consult the solution. Then he should try the problem again without the solution at hand. If the problems are worked through systematically using the solutions as little as possible each time, the student should gain confidence and be in a position to attempt those from textbooks for which there is often no solution available.

The problems in this volume are suitable for students of engineering, physics or mathematics who are undergoing a second course of field theory.

As in the first volume, the authors have attempted to acknowledge the sources of all problems where applicable. If any have been omitted it is entirely unintentional and we can only offer our apologies.

K. Foster
R. Anderson

CONTENTS

ACKNOWLEDGEMENTS

The following abbreviations have been used in denoting the source of the various problems:

U.A. – University of Aston
U.D. – University of Durham
U.E. – University of Edinburgh
U.Ld. – University of Leeds
U.L. – University of London
U.M. – University of Manchester
U.S. – University of Sheffield
U.Shn. – University of Southampton
U.St. – University of Strathclyde
U.St.A. – University of St. Andrews
L.C. – Lanchester College of Technology, Coventry

The authors wish to thank the Senates of the above Universities for permission to reproduce examples from their examination papers. The bodies concerned are, of course, in no way committed to approval of the solutions given in this book.

PART I

PROBLEMS

CHAPTER 1

MAXWELL'S EQUATIONS

Introduction

The five basic equations of electromagnetic theory are:

$$\nabla \times \mathbf{E} = -\frac{\partial \mathbf{B}}{\partial t} \qquad \ldots.(1)$$

$$\nabla \times \mathbf{H} = \mathbf{J} + \frac{\partial \mathbf{D}}{\partial t} \qquad \ldots.(2)$$

$$\nabla \cdot \mathbf{D} = \rho \qquad \ldots.(3)$$

$$\nabla \cdot \mathbf{B} = 0 \qquad \ldots.(4)$$

$$\mathbf{F} = q(\mathbf{E} + \mathbf{v} \times \mathbf{B}) \qquad \ldots.(5)$$

Equations (1) to (4) are known collectively as Maxwell's equations and relate the field vectors, the charges and the current densities at any ordinary point in a stationary medium–the sort of situation we shall be dealing with in this chapter. We note in passing that these linear, partial differential equations of the first order are perfectly general and we can if we wish convert them into an integral formulation by the well known theorems of vector calculus due to Gauss and Stokes. This latter presentation can be useful for problems with a high degree of symmetry, or, in other cases, for a deeper physical insight into the problem. Clearly both give the same information and in many of the problems encountered it is simply a question of convenience and personal taste which method one employs.

Equation (5) is the Lorentz relation which in essence defines our field vectors **E** and **B**.

In any particular problem we also require the constitutive equations, $\mathbf{D} = \epsilon_0\epsilon_r\ \mathbf{E}$, $\mathbf{B} = \mu_0\mu_r\ \mathbf{H}$ and $\mathbf{J} = \sigma\mathbf{E}$ where ϵ_r, μ_r and σ must be obtained from experiment. They can in general be quite complicated parameters but we will only consider the simplest cases.

The above equations then, together with the conditions imposed by the boundaries, are all that we require in principle to solve any electromagnetic theory problem.

Maxwell's equations are a set of mixed equations, i.e. more than one field vector appears in two of the equations, and in order to make them easier to solve several artifices are available for transforming them into equivalent relations in terms of a single variable—the price we pay for this is to increase the order of the equations from one to two. If, for example, we consider a region of free space in which there are no free charges, and take the curl of each side of (1) we obtain, by substituting from (2)

$$\nabla \times \nabla \times \mathbf{E} = -\mu_0 \frac{\partial}{\partial t}\left[\epsilon_0 \frac{\partial \mathbf{E}}{\partial t}\right] = \nabla(\nabla \cdot \mathbf{E}) - \nabla^2 \mathbf{E}$$

hence

$$\nabla^2 \mathbf{E} - \mu_0 \epsilon_0 \frac{\partial^2 \mathbf{E}}{\partial t^2} = 0$$

and a similar equation in **H**, i.e.

$$\nabla^2 \mathbf{H} - \mu_0 \epsilon_0 \frac{\partial^2 \mathbf{H}}{\partial t^2} = 0$$

—two second order equations.

The concept of potential is another way of separating Maxwell's equations whereby we introduce new field quantities which automatically satisfy some of these equations. For example, since the divergence of **B** is always zero, we can introduce a new field vector **A** such that $\mathbf{B} = \nabla \times \mathbf{A}$. If this is substituted into (1) and the terms rearranged we find that

$$\nabla \times \left(\mathbf{E} + \frac{\partial \mathbf{A}}{\partial t}\right) = 0$$

thus $[\mathbf{E} + (\partial \mathbf{A}/\partial t)]$ can be represented by the gradient of some scalar field $\phi \ldots -\nabla\phi$. Hence

$$\mathbf{B} = \nabla \times \mathbf{A}$$

$$\mathbf{E} = -\nabla\phi - \frac{\partial \mathbf{A}}{\partial t}$$

A is called the vector magnetic potential and ϕ the scalar electric potential.

By substituting these expressions back into Maxwell's equations and imposing the Lorentz condition on **A** and ϕ, i.e.

$$\nabla \cdot \mathbf{A} + \mu_0\epsilon_0\frac{\partial\phi}{\partial t} = 0$$

we obtain two inhomogeneous wave equations,

$$\nabla^2\mathbf{A} - \mu_0\epsilon_0\frac{\partial^2\mathbf{A}}{\partial t^2} = -\mu_0\mathbf{J}$$

$$\nabla^2\phi - \mu_0\epsilon_0\frac{\partial^2\phi}{\partial t^2} = -\frac{\rho}{\epsilon_0}$$

where again for simplicity we have assumed that $\epsilon_r = \mu_r = 1$.

These equations can be solved and lead to the 'retarded potentials'

$$\phi = \frac{1}{4\pi\epsilon_0}\int_v \frac{[\rho]}{r}\,\mathrm{d}v, \quad \mathbf{A} = \frac{\mu_0}{4\pi}\int_v \frac{[\mathbf{J}]}{r}\,\mathrm{d}v$$

The square brackets have the usual meaning in this context that the contribution to the potential at some instant of time t due to a source distant r from the field point is governed not by the value of the source at time t but by its value at the earlier time $[t - (r/c)]$ where c is a characteristic velocity:

$$\frac{1}{\sqrt{\mu_0\epsilon_0}}$$

This method can be very useful when the sources are known and it is desired to calculate the fields.

By the same arguments that are used in determining the retarded potentials, we can arrive at a representation in terms of 'advanced potentials' which are equally valid in a mathematical sense. In practice we usually ignore such solutions as they seem to violate our notions regarding causality.

There are other potential functions than those introduced here but they are not necessary for the solution of the problems in this chapter.

So far we have said nothing about time variations. The simplest situation is to have none, in which case we just put $\partial/\partial t$ equal to zero everywhere. Maxwell's equations separate into two sets, one set for the electric field and the other for the magnetic field.

A useful time dependence to employ is $e^{j\omega t}$. This harmonic variation is extremely important in many engineering problems as well as leading to great simplification in the mathematics. Formally we simply replace $\partial/\partial t$ by $j\omega$. Furthermore, the method is not as limited as at first sight it may appear since we can always make a Fourier analysis of any practical time dependence and use superposition.

Finally a few words about Poynting's Theorem. This most important theorem, both in its real and complex form, is used throughout the book and it is assumed that the reader is familiar with it and appreciates its interpretation. If not, several excellent discussions may be found in the reference given at the end of the book.

Problems

1.1. Explain what is meant by a retarded potential in electrodynamics.

An infinite uniform wire occupies the z axis in free space, the current in it is vanishing for $t \leqslant 0$ and having the constant value J for $t > 0$. Show that the vector potential at the point P, whose cylindrical coordinates are r, ϕ, z is given by

$$\mathbf{A} = \frac{\mu_0 J}{2\pi}\,\mathbf{k}\log\left\{\frac{ct + \sqrt{(c^2t^2 - r^2)}}{r}\right\} \quad \text{for } t \geqslant \frac{r}{c}$$

$$= 0 \quad \text{for } t < \frac{r}{c}$$

and that the electromagnetic field of the current at P when $t \geqslant (r/c)$ is given by

$$\mathbf{E} = -\frac{\mu_0 c J\,\mathbf{k}}{2\pi\sqrt{(c^2t^2 - r^2)}}$$

$$\mathbf{B} = \frac{\mu_0 J c t\,\hat{\boldsymbol{\phi}}}{2\pi r\sqrt{(c^2t^2 - r^2)}}$$

Find the flux of energy across unit length of a cylinder passing through P and coaxial with the wire, when $t \geqslant (r/c)$ (U.L.)

1.2. Obtain the retarded potential solution to the equation

$$\nabla^2\mathbf{A} - \frac{1}{c^2}\frac{\partial^2\mathbf{A}}{\partial t^2} = -\mu_0\,\mathbf{j}$$

where **A** is the vector potential of an electromagnetic field in free space whose source is a current distribution of density **j**.

A current I cos ωt flows in a thin wire bent in the form of a circle of radius a. If $\omega a/c \ll 1$, show that at points in the plane of the circle whose distances R from the centre of the circle are large compared with a, the component of **A** perpendicular to the radius vector is approximately

$$\frac{\mu_0 a^2 I}{4R}\left\{\frac{1}{R}\cos\omega\left(t-\frac{R}{c}\right)-\frac{\omega}{c}\sin\omega\left(t-\frac{R}{c}\right)\right\} \qquad \text{(U.L.)}$$

1.3. Write down Maxwell's equations for the electromagnetic field vectors **E** and **H** in empty space. If V is a solution of

$$\nabla^2 V = \frac{1}{c^2}\frac{\partial^2 V}{\partial t^2}$$

show that

$$\mathbf{E} = \nabla\left(\frac{\partial V}{\partial z}\right) - \mathbf{k}\frac{1}{c^2}\frac{\partial^2 V}{\partial t^2}$$

where **k** is a unit vector in the z direction, is a possible form of **E**. Find the corresponding form of **H**.

Given that

$$\nabla^2 V \equiv \frac{1}{r}\frac{\partial^2}{\partial r^2}(rV)$$

show that

$$V(r,t) = \frac{1}{r}\sin\frac{r}{a}\sin\frac{ct}{a}$$

satisfies the equation for V.

Show also that

$$\mathbf{E} = \mathbf{k}\left(\frac{1}{r}\frac{\partial V}{\partial r}+\frac{V}{a^2}\right)+\mathbf{r}\frac{z}{r}\frac{\partial}{dr}\left(\frac{1}{r}\frac{\partial V}{\partial r}\right) \qquad \text{(U.L.)}$$

1.4. Write down Maxwell's equations for free space and show how any field satisfying these equations may be expressed in terms of a scalar and vector potential ϕ and A.

If $u = t - r/c$, $r^2 = x^2 + y^2 + z^2$ and k is a unit vector in the z direction and $f(u)$ is a function of u, show that

$$\mathbf{A} = \mathbf{k}\frac{f'(u)}{cr}$$

is a possible vector potential and find a form for a related scalar potential.

(U.L.)

1.5. A surface current of density i flows in a metal sheet occupying the plane $z = 0$ and the rest of space is empty.

By integrating Maxwell's equation for curl **H** over a suitable surface, or otherwise, express i in terms of the discontinuity in the magnetic field at $z = 0$. State the other conditions which must hold.

Verify that a possible solution of Maxwell's equations is, using Cartesian coordinates,

$$\mathbf{E} = [0, 0, A \sin n(x - ct)]$$

$$\mathbf{H} = [0, -\lambda\, 4 \sin n(x - ct), 0]$$

for $z > 0$, where λ is a constant depending on the units, and $\mathbf{E} = \mathbf{H} = 0$ for $z < 0$.

Determine the current in the sheet necessary to maintain the oscillation.

(U.L.)

1.6. The magnetic vector potential, A, can be shown to satisfy the differential equation

$$\nabla^2\mathbf{A} - \frac{1}{c^2}\frac{\partial^2\mathbf{A}}{\partial t^2} = -\mu\mathbf{J}$$

Show, by analogy with Poisson's equation in electrostatics, that for the case of no time variation of A,

$$\mathbf{A} = \frac{\mu}{4\pi}\int\frac{\mathbf{J}\,\mathrm{d}V}{r}$$

where the integral is taken throughout the volume carrying the current.

A wire of small cross-section coincides with the y-axis from $y = 0$ to $y = l$ and a similar wire is placed parallel to this at $x = a$ and again extends from $y = 0$ to $y = l$.

Show that the vector potential at the first wire (at $x = 0, y = y_0$) due to a current I flowing in the second wire is given by

$$A(y_0) = \frac{\mu I}{4\pi}\left\{\sinh^{-1}\left(\frac{y_0}{a}\right) - \sinh^{-1}\left(\frac{y_0 - l}{a}\right)\right\}$$

Explain how this expression could be used to obtain the mutual inductance between the two wires.

[Note: $\displaystyle\int \frac{dx}{\sqrt{(a^2 + x^2)}} = \sinh^{-1}\left(\frac{x}{a}\right)$] (U.E.)

1.7. Starting from Maxwell's equations derive an equation which demonstrates the significance of the quantity $\oint \mathbf{E} \times \mathbf{H} \,.\, d\mathbf{S}$ where **E** and **H** are values of the electric and magnetic fields on a closed surface **S**.

A parallel plate capacitor consists of two discs of diameter d spaced distance h apart with voltage V between them. The plate spacing is then increased at a uniform rate for one second. The spacing is then $2h$.

Find the induced magnetic field and the Poynting Vector at the edge of the capacitor plates as the plates are separated and use Poynting's Theorem to relate the change in stored energy to the energy lost as the plates are separated. List any assumptions made. (U.S.)

1.8. When there is no change of any quantity with time, two of Maxwell's equations become

$$\text{curl}\,\mathbf{E} = 0$$
$$\text{div}\,\mathbf{D} = \rho$$

Show that this *pair* of equations can be reduced to a *single* equation with the help of a scalar potential.

Write down the corresponding pair of Maxwell's equations involving the magnetic field under static conditions, and show how a potential function can again be used to reduce them to a single equation.

By analogy with the electrostatic case, derive a formula for this potential function in terms of the current density. (U.S.)

1.9. Show that the potential function $\phi = A(1 - z/L) \ln (r/b)$, where A is an arbitrary constant, satisfies the Laplace equation in the cylindrical coordinates (r, θ, z)

$$\frac{1}{r}\frac{\partial}{\partial r}\left(r\frac{\partial\phi}{\partial r}\right) + \frac{1}{r^2}\frac{\partial^2\phi}{\partial\theta^2} + \frac{\partial^2\phi}{\partial z^2} = 0$$

in a region bounded by perfect conductors at $r = b$ and at $z = L$.

Hence, or otherwise, if a voltage V is applied to the coaxial resistor of *Figure 1.1* (in such a fashion that the potential across the input surface

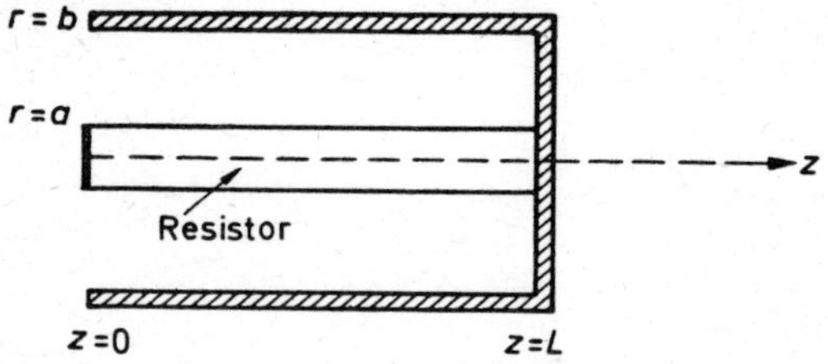

Figure 1.1.

varies logarithmically), calculate the electric and magnetic fields in the non-conducting region.

By means of the Poynting vector show that the power flow into the resistive inner cylinder is the same as that computed from conventional circuit analysis. (L.C.)

1.10. If the fields in Maxwell's equations all have a sinusoidal time dependence such as

$$\mathbf{E} = \mathbf{E}_0 \cos(\omega t + \phi_1) \ldots \text{etc.}$$

observe that they can be written in the forms equivalent to

$$\mathbf{E} = \tfrac{1}{2}[\hat{\mathbf{E}} \exp(\mathrm{j}\omega t) + \hat{\mathbf{E}}^* \exp(-\mathrm{j}\omega t)]$$

where

$$\hat{\mathbf{E}} = \mathbf{E}_0 \exp(\mathrm{j}\phi_1)$$

Noting that Maxwell's equations are linear and hence that the principle of superposition can be applied, derive the *phasor* Maxwell equations relating the *phasor vectors* $\hat{\mathbf{E}}$, etc.

Starting from the expressions for the stored electric and magnetic energy densities and for the Poynting vector, derive corresponding expressions for the time average stored energy densities and the time average power flow vector (complex Poynting vector) in terms of the phasor vectors.

CHAPTER 2

STATIC FIELDS

Introduction

The central problem of electrostatics is the solution of Laplace's equation, or Poisson's equation, under various imposed boundary conditions. In this chapter we study such problems in a variety of coordinate systems–these being chosen so that the partial differential equation $\nabla^2\phi = 0$ admits of a 'separation of variables' type of solution. We begin then by assuming that such a solution exists. In general there are an infinite number of possible solutions and we select the correct one by applying the given boundary conditions. The uniqueness theorem tells us that since we have chosen the solution that satisfies all the boundary conditions then it holds at all interior points and is the only one. At this stage the problem is basically considered solved.

As far as electromagnetic theory appreciation and physical content is concerned, the reader need have little more knowledge than that required to do problems of the standard of those in Volume 1; however it is assumed that he is now mathematically more mature. The mathematics usually required involves a little vector calculus, Bessel functions, Legendre polynomials, Fourier series and, in Problem 2.3, a Fourier-Bessel expansion. Even so, the reader should not feel that he has to know all about these; rather it is suggested that he keeps a good maths book handy as he is attempting the solutions and that he refers to it as often as necessary. The desire to solve a particular problem can be a powerful stimulus to brush up on the mathematics as one goes along. There is little more to be said than that practice is required.

Problems

2.1. *Figure 2.1* shows a long thin strip of resistive material of width a. The electric potential, ϕ, satisfies Laplace's equation on the strip and is subjected to the following conditions:

(*i*) $\phi \to 0$ as $y \to \infty$

(*ii*) $\phi = 0$, where $x = 0$, and $x = a$,

(*iii*) $\phi = \begin{cases} x & 0 \leqslant x \leqslant a/2 \\ (a - x) & a/2 \leqslant x \leqslant a \end{cases}$ where $y = 0, 0 \leqslant x \leqslant a$

(*a*) By separating the variables, solve for ϕ on the strip.

(*b*) If the surface resistivity of the strip is 1000 Ω/square, estimate the current in a strip of width 0·1 m.

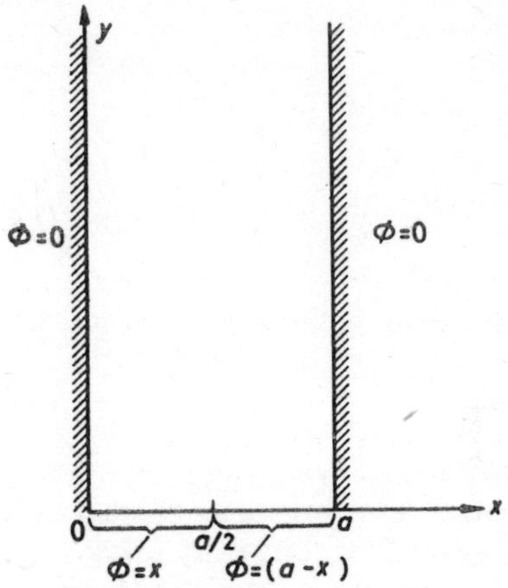

Figure 2.1.

2.2. Show that a solution of Laplace's equation

$$\frac{\partial^2 V}{\partial x^2} + \frac{\partial^2 V}{\partial y^2} = 0$$

for which $V = 0$ when $x = 0$ and when $y = 0$; $V = V_0$ when $x = a$; and $(\partial V/\partial y) = 0$ when $y = b$, can be written in the form

$$V = \frac{4V_0}{\pi}\sum_{n=0}^{\infty} \frac{\sinh\{(2n+1)\pi x/2b\}\sin\{(2n+1)\pi y/2b\}}{(2n+1)\sinh\{(2n+1)\pi a/2b\}} \qquad \text{(U.E.)}$$

2.3. Show that the equation

$$\frac{\partial^2 V}{\partial r^2} + \frac{1}{r}\frac{\partial V}{\partial r} + \frac{\partial^2 V}{\partial z^2} = 0$$

has solutions of the form

$$V = \{AJ_0(pr) + BY_0(pr)\}\{C\cosh pz + D\sinh pz\}$$

If V satisfies the conditions

(*i*) V is finite as $r \to 0$, $(0 < z < h)$
(*ii*) $V = 0$ at $r = a$, $(0 < z < h)$
(*iii*) $V = 0$ at $z = 0$, $(0 < r < a)$
(*iv*) $V = E\,(a^2 - r^2)$ at $z = h$, $(0 < r < a)$

where E is a constant, show that

$$V = 8a^2E \sum_{i=1}^{\infty} \frac{1}{\lambda_i^3 J_1(\lambda_i)} \frac{\sinh(\lambda_i z/a)}{\sinh(\lambda_i h/a)} J_0\left(\frac{\lambda_i r}{a}\right)$$

where λ_i is the i-th positive root of the equation $J_0(\lambda) = 0$. (U.E.)

2.4. Show that if the two dimensional form of Laplace's equation is expressed in terms of the plane polar coordinates r and θ, it becomes

$$\frac{\partial^2\phi}{\partial r^2} + \frac{1}{r}\frac{\partial\phi}{\partial r} + \frac{1}{r^2}\frac{\partial\phi}{\partial\theta^2} = 0$$

The axis of an infinitely long hollow circular cylinder of conducting material is taken as the axis of cylindrical polar coordinates (r, θ, z), and the inner surface, of radius a, is given an electric potential E, when $0 < \theta < \pi$, and $-E$ when $-\pi < \theta < 0$.

Show that the potential within the cylinder is given by

$$\phi = \frac{4E}{\pi} \sum_{n=0}^{\infty} \left(\frac{r}{a}\right)^{2n+1} \frac{\sin(2n+1)\theta}{(2n+1)} \qquad \text{(U.D.)}$$

2.5. An infinitely long hollow iron cylinder of permeability μ, the cross section being concentric circles of radii a, b $(a < b)$, is placed in a uniform field of magnetic force the direction of which is perpendicular to the generators of the cylinder. Show that the flux of induction B through the space occupied by the cylinder is changed by inserting the cylinder in the field, in the ratio

$$b^2(\mu + 1)^2 - a^2(\mu - 1)^2 : 2\mu\{b^2(\mu + 1) - a^2(\mu - 1)\} \qquad \text{(U.E.)}$$

2.6. Prove that the mechanical force on the surface of a charged conductor *in vacuo* is $\sigma^2/2\epsilon_0$ per unit area, where σ is the surface charge density.

A thin spherical earthed conducting shell of radius a is placed in a uniform electric field of intensity **E**. If the shell be cut into two halves by a plane through the centre parallel to **E**, show that each half will be acted on by a force $4\pi\epsilon_0\ [(9/32)\,a^2E^2]$. (U.D.)

2.7. *Figure 2.2* shows a hollow dielectric shell of relative permittivity K placed in what was, before the insertion of the shell, a uniform electric field of intensity **E**. Investigate the field in the empty interior region.

If a point charge q is now placed at the centre of the sphere, calculate (a) the force on the point charge, (b) the force on the dielectric shell.

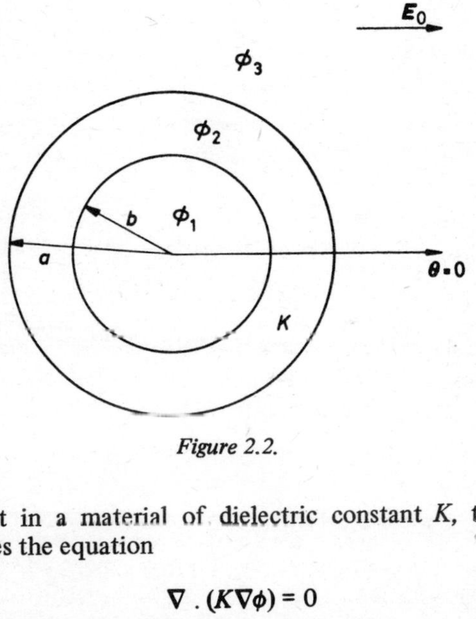

Figure 2.2.

2.8. Prove that in a material of dielectric constant K, the electrostatic potential satisfies the equation

$$\nabla\,.\,(K\nabla\phi) = 0$$

A spherical shell, bounded by two concentric spheres of radii λa, a ($\lambda < 1$), is made of material whose dielectric constant at a point distant r from the centre of the spheres is given by $K = (a/r)^2$. The shell is placed in a uniform electric field of intensity E. Show that in the shell ($\lambda a \leqslant r \leqslant a$) the potential is of the form $f(r)\cos\theta$ where

$$\frac{d^2f}{dr^2} = \frac{2f}{r^2}$$

Hence show that the field inside the shell ($r < \lambda a$) is uniform and of magnitude

$$\frac{9E\lambda}{4 + 4\lambda^2 + 2\lambda^3 - \lambda^5} \qquad \text{(U.S.)}$$

2.9. Show that the potential outside a uniformly charged ring of total charge Q and radius c is given by

$$\left.\begin{aligned} V &= \frac{Q}{4\pi\epsilon_0 c}\sum_{n=0}^{\infty}\left(\frac{c}{r}\right)^{n+1} P_n(0)\, P_n(\cos\theta) \quad \text{for } r > c \\ V &= \frac{Q}{4\pi\epsilon_0 c}\sum_{n=0}^{\infty}\left(\frac{r}{c}\right)^{n} P_n(0)\, P_n(\cos\theta) \quad \text{for } r < c \end{aligned}\right\} \begin{aligned}&\text{and for}\\ &r = c\\ &\text{when}\\ &\theta \neq \frac{\pi}{2}\end{aligned}$$

where (r, θ, z) are cylindrical polar coordinates referred to the centre of the circle as origin, with the z axis coinciding with the axis of the circle and $P_n(\mu)$ are Legendre polynomials

$$P_0(\mu) = 1,\quad P_1(\mu) = \mu,\quad P_2(\mu) = \tfrac{1}{2}(3\mu^2 - 1)\ \ldots \text{etc.}$$

Two concentric hollow conducting spheres of radii a and b are maintained at zero potential. A thin uniformly charged ring of total charge Q and radius c ($b > c > a$) is placed with its centre at the centre of the spheres. Show that the charge induced on the inner sphere is

$$-Q\frac{a\,(b-c)}{c\,(b-a)} \qquad \text{(U.D.)}$$

2.10. A ring of radius c has a total charge e distributed uniformly round its circumference, and is situated in an infinite uncharged medium of s.i.c. K. There is a spherical cavity in the medium, of radius $a < c$, and its centre O coincides with the centre of the ring. Prove that the potential at a point P in the cavity is

$$\frac{e}{4\pi\epsilon_0 c}\sum_{n=0}^{\infty}\frac{4n+1}{K(2n+1)+2n}\cdot\frac{(-1)^n(2n)!}{2^{2n}(n!)^2}\left(\frac{r}{c}\right)^{2n} P_{2n}(\cos\theta)$$

where r is the distance OP, and θ is the angle between OP and the normal at O to the plane of the ring. (U.D.)

2.11. A hoop of radius b carrying a uniformly distributed charge $-e$ is placed with its centre at $z = b$ and its axis along the z axis. Another similar hoop carrying a charge $+e$ is similarly placed at $z = -b$. Show that in the neighbourhood of the origin the potential due to the hoops has the form

$$\phi = \frac{e}{4\pi\epsilon_0 b\sqrt{2}}\left[-\frac{r}{b}P_1(\cos\theta) + \frac{r^3}{8b^3}P_3(\cos\theta) + \ldots\right]$$

If a small earthed sphere of radius a is placed at the origin, show that the electric intensity at the centre of the first hoop is increased by

$$\frac{ea^3\sqrt{2}}{4\pi\epsilon_0 b^5}\left[1 + 0\left(\frac{a^4}{b^4}\right)\right]$$ (U.E.)

2.12. A point charge e is placed at a distance d from the centre of a spherical cavity of radius a ($> d$) inside a uniform dielectric medium whose dielectric constant is ϵ. Calculate the potential inside the cavity and hence show that the charge experiences a force

$$\frac{(\epsilon - 1)e^2}{4\pi\epsilon_0 ad}\sum \frac{l(l+1)}{\epsilon(l+1) + l}\left(\frac{d}{a}\right)^{2l}$$

away from the centre. (U.E.)

2.13. Two circular wires of radii a and b have a common centre and are free to turn on an insulating axis which is a diameter to both. Show that when the wires carry currents I and I' and $b \ll a$ a couple of magnitude

$$\frac{\mu_0\pi b^2}{2a}\left[1 - \frac{9}{16}\frac{b^2}{a^2} + 0\left(\frac{b^4}{a^4}\right)\right]II'$$

is required to hold them with their planes at right angles. (U.E.)

2.14. A surface current in the plane $y = 0$ has the form

$$K_z = K_1 \cos\frac{2\pi x}{L}$$

A similar surface current in the plane $y = d$ has the form

$$K'_z = K_2 \sin \frac{2\pi}{L}(x + l)$$

Calculate the x and y components of the force per periodic length L, and per unit length in the z direction, between the two current distributions.

Hence show that the magnitude of the force vector is independent of l and makes an angle $2\pi l/L$ with the plane $y = 0$. (U.S.)

CHAPTER 3

CONFORMAL TRANSFORMATION AND VARIATIONAL TECHNIQUE

Introduction

The problems in this chapter are examples of two of the most powerful methods of solving electromagnetic problems. Naturally at the level of this book they are relatively elementary examples of these methods but they should give the student an appreciation of the power of the analysis together with a certain amount of confidence in the application.

It has been decided to restrict the examples of conformal transformation to those which are basically practical and those which do not require a detailed knowledge of elliptic and associated functions for their solution. Problems **3.4** and **3.5** are not exceptions to this rule if we accept that the configuration of *Figure 3.14* (the upper half-space) is standard with a capacitance of the form given in Problem **3.4** in terms of elliptic integrals. The student does not have to use such integrals and, in fact, for the purpose of solution the capacitance could equally well be taken to be that deduced from the approximation given in the hint to Problem **3.4**. This particular w-plane configuration is an extremely important one and the student should be introduced to it even without direct knowledge of its derivation, which does involve elliptic functions. In books mentioned under References (p. 239), there are many problems on some of the other aspects of transformation and the student should also attempt these.

On the variational side the examples have been restricted to electrostatics. In this case the simple variational solution is often one of minimum energy and as a result an upper bound to the capacitance is often all that is found. The student should appreciate that a lower bound can also be found although the theory for this is considered to be beyond the level assumed in this book (see Reference 8). Unfortunately very few texts consider the variational problem at student level so that the student is advised to consult Reference 3 in which there is a very readable chapter covering the basic aspects required for these problems. Problem **3.9** is a rather general theorem and **3.10** is an application of this, using also the result of **3.1**, to obtain both an upper and a lower bound to capacitance.

Perhaps the most striking feature that the student will observe as he works through the examples is the remarkable accuracy obtained with the crudest of approximations in the guess of the potential functions. In the solutions the linear potential function is used for simplicity but the student must realise that a more sophisticated function should be used in practice. As an indication of our meaning, we know that at the edge of a conductor the field will be (theoretically) infinite. This means that the function chosen for potential should have an infinite slope at this point, a property never obeyed by the linear function. We would suggest that the interested student try a modified potential function on some of the examples in an attempt to obtain a better answer. Remember that it is better if it is lower! A little practice at this soon gives the feeling for 'a judicious choice'.

Finally we would address a few words to the student who would prefer exact answers. Many practical problems of field theory are insoluble in exact terms and many more have solutions which are infinite series. Although the latter may be able to be computed to as high a degree of accuracy as anyone would wish, this may be an expensive process. In practical terms a real accuracy of much better than 1% is often unusable and very often an accuracy of about 5% is adequate. Thus, although we do not wish to persuade the student not to seek an exact solution, we would stress that he must develop an appreciation of approximation. To this end he is asked to obtain approximate solutions in some cases.

Problems

3.1. Two parallel, earthed, conducting plates are spaced apart by a distance b. A line charge of density σ C/m is placed equidistant from the plates and parallel to them. Determine the potential in the space between the plates and discuss, in general terms, the form of the equipotential surfaces.

3.2. Suppose that, in the configuration of the previous problem the line charge was not equidistant from the plates but a distance $(\alpha b/2)$ from the centre, where $0 < \alpha < 1$. Determine the potential between the plates in this case.

If a second line charge of equal strength is now placed at the image distance $(-\alpha b/2)$ from the centre what is the new potential function?

3.3. Two parallel conducting plates are spaced apart by a distance b. A conducting septum projects from one plate to a distance a from the other as shown in *Figure 3.1.* If the plate with the septum is held at a potential V relative to the other plate, determine

(*a*) the potential in the plane of the gap, and

(*b*) the excess capacitance (that is the capacitance over and above the normal parallel plate capacitance) due to the presence of the septum.

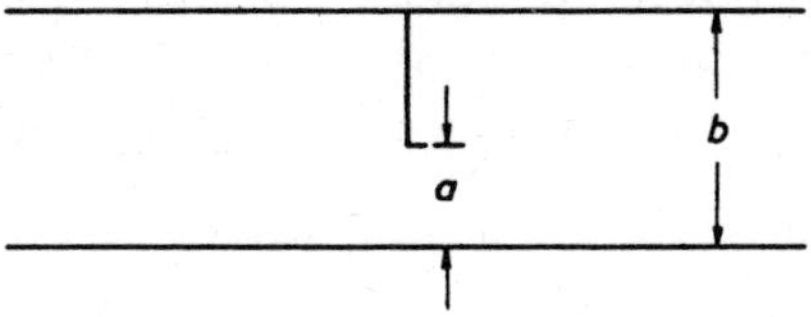

Figure 3.1.

3.4. A strip transmission line is formed from two earthed parallel conducting plates a distance b apart with a centrally disposed strip of width a

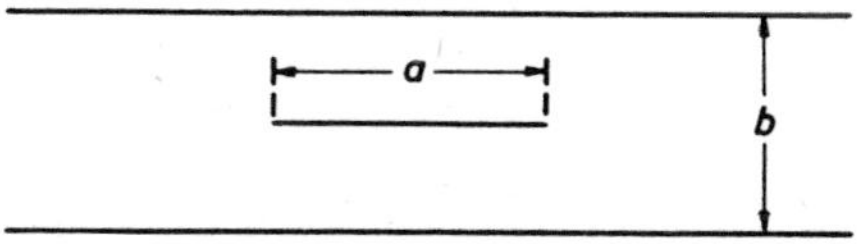

Figure 3.2.

as shown in *Figure 3.2.* Obtain an expression for the capacitance per unit length of this transmission line in the form

$$C = 4\epsilon K'(k)/K(k) \quad \text{where} \quad k = \text{sech}\,(\pi a/2b)$$

and where $K'(k)$ and $K(k)$ are the usual complete elliptic integrals.

Show that this can be expressed approximately as the sum of a parallel plate value (of width a) and an 'edge capacitance'.

[Hint: for small k the ratio $K'/K \approx (2/\pi) \ln (4/k)$]

3.5. Two strip transmission lines similar to that of the previous problem are coupled together by placing their inner conductors (strips of width a) inside the same ground planes and a distance s apart, as shown in *Figure 3.3(a)*. Viewed as a three-wire structure the various capacitances per unit length can be defined as shown in *Figure 3.3(b)*. Obtain expressions for C_p and C_c.

(Hint: consider the two cases when the inner conductors are first at the same potential and secondly at equal but opposite potentials.)

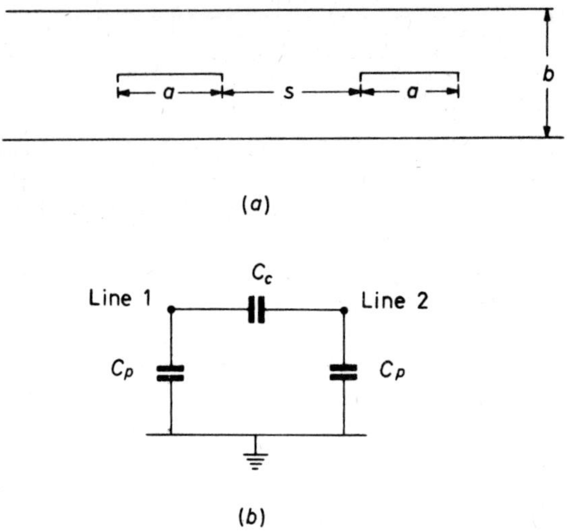

Figure 3.3.

3.6. Two concentric, conducting spheres have radii a and b where $a < b$. Assuming in the first instance that the potential varies linearly with the radial coordinate, obtain, by variational methods, an upper bound for the capacitance of the two spheres. Compare the value obtained with the correct value

$$C = 4\pi\epsilon \frac{ab}{(b - a)}$$

Repeat the exercise for a quadratic variation of potential.

3.7. Consider the strip transmission line of *Figure 3.2.* By assuming a linear potential variation from the ground planes to the edge of the strip and using the variational method, obtain an upper bound to the capacitance per unit length. Compare the answer obtained with that from Problem 3.4.

3.8. A square cross-section coaxial line has the dimensions shown in *Figure 3.4.* By the variational technique calculate an upper bound to the capacitance per unit length. (Note: The solution of this problem by conformal transformation is beyond the scope of this book but for comparison purposes the exact answer is 90·4 pF/m).

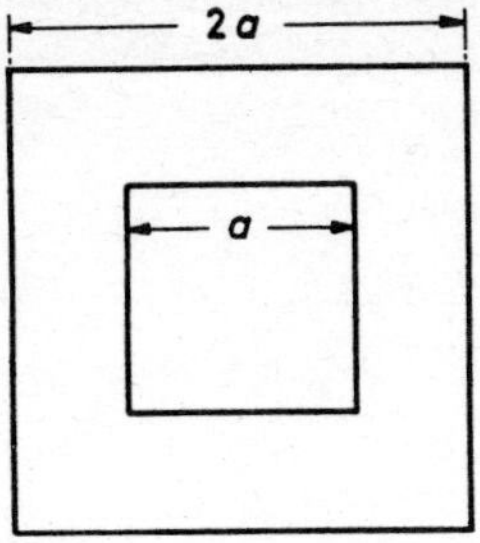

Figure 3.4.

3.9. Consider three surfaces S_1, S_2 and S_3 as shown in *Figure 3.5* in which S_3 encloses S_2 and S_2 in turn encloses S_1. If S_2 and S_3 are replaced by con-

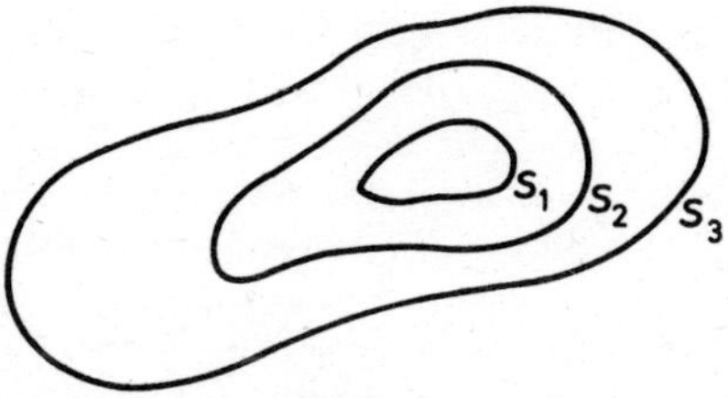

Figure 3.5.

ductors a capacitance C_{23} is obtained. If, as a separate exercise, S_1 and S_3 are replaced by conductors a capacitance C_{13} is obtained. Show that $C_{23} > C_{13}$.

3.10. A transmission line consists of a circular inner conductor of radius a centrally disposed between two ground planes of spacing b as shown in *Figure 3.6.* Use the result of Problem 3.9 together with the field of Problem 3.1 to obtain both an upper and a lower bound to the capacitance per unit length of this line.

Plot the results obtained and assess the accuracy of an estimate of this capacitance.

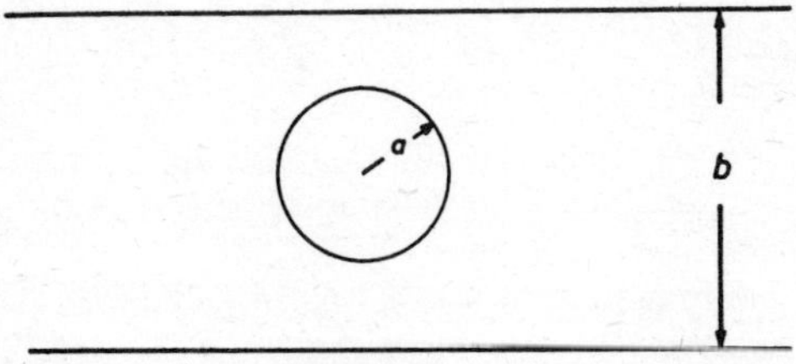

Figure 3.6.

CHAPTER 4

WAVES

Introduction

In this section we construct solutions of Maxwell's equations that possess wave-like properties and, for mathematical simplicity, consider media for which the permeability and permittivity are scalar quantities.

The first two problems involve vector manipulation of Maxwell's equations in order to arrive at the wave equations; the first for free space and the second for a lossy medium. These highlight the important role played by the displacement current term in radiation phenomena for without it there could be no propagation in free space. Also we observe that for a conducting medium, the Joule heating causes the wave to be exponentially attenuated with distance.

In the problems that follow we only consider plane waves varying harmonically in time–again for mathematical simplicity. This is not quite so restricting as it appears at first sight since the properties of, say, spherical or cylindrical waves approximate closely those of plane waves at large distances from their sources. By studying these simpler waves, then, we can learn a good deal about the more complicated ones.

Now just because an **E** and **H** that satisfy Maxwell's equations automatically satisfy the wave equations, this does not imply that the converse is true–in fact it is not so. However we always avoid this dilemma in practice by solving the wave equation for **E**, for instance, and then, substituting this expression for **E** in the curl **E** relation, derive an expression for **H** (or vice-versa). There is then no doubt that we have an admissible solution. Furthermore, it is convenient although not essential to work in terms of **E** and **H** directly rather than by introducing potential functions as we did in Chapter 1; this is not necessarily so in more complicated wave types.

Problem 4.3 involves an appreciation of the Brewster angle and in the solution we have expressed the equation in matrix form. True, in this problem there is little or nothing to be gained by doing so, but in more complicated problems involving multiple dielectric interfaces the authors find it superior; as we hope later problems justify this point of view. Whereas matrix methods are widely used in engineering circles, this is not so in other disciplines, students of which need not be handicapped as the method in itself does not advance our understanding of the electromagnetic theory involved. It is simply a question of convenience and personal taste.

Problem **4.4** is concerned with the phase and group velocities of a plane wave in a lossy medium, an excellent account of which is given in Reference 1.

The remaining examples all involve multiple boundaries and are simple boundary value problems chosen with engineering applications in mind. Again matrix methods have been employed. Problems **4.8** and **4.9** illustrate the close connection between plane waves in space and uniform transmission line theory (see Reference 2). The Smith Chart solution to **4.9** in particular shows an engineering method worth noting–especially in the solution of numerical problems.

Problems

4.1. Assume free space conditions ($\rho = 0$, $\mathbf{J} = 0$) and express Maxwell's equations in terms of **H** and **E** only. Take the curl of the point form of Ampère's law, and using the vector identity $\nabla \times \nabla \times \mathbf{A} \equiv \nabla(\nabla . \mathbf{A}) - \nabla^2 \mathbf{A}$ where **A** is any differentiable vector, show that

$$\nabla^2 \mathbf{H} = \mu_0 \epsilon_0 \frac{\partial^2 \mathbf{H}}{\partial t^2}$$

Obtain an analogous expression for **E** by taking the curl of the point form of Faraday's law. (U.St.)

4.2. The wave equation for a time-varying electromagnetic field can be expressed in terms of the magnetic component:

$$\nabla^2 \mathbf{H} = \mu\sigma \frac{\partial \mathbf{H}}{\partial t} + \mu\epsilon \frac{\partial^2 \mathbf{H}}{\partial t^2}$$

Derive the expression for the propagation constant for a plane wave travelling in a 'good conductor', explaining what is meant by this expression.

At 1 MHz a plane wave is incident normally on the surface of sea water ($\sigma = 4$, $\epsilon_r = 80$). Find the depth at which the magnetic field is one per cent of its value at the surface. (U.E.)

4.3. A plane e.m. wave in free space, polarized in the plane of incidence, impinges upon the plane boundary of a thick dielectric at an angle of 56·6° to the normal to that boundary. If there is no reflected wave, what is the value of ϵ_r for the dielectric? Derive all expressions used.

Discuss in qualitative terms the polarizations of the reflected and transmitted waves, if by some means the **E** vector of the incident wave is rotated through a small angle. (U.M.)

4.4. Define phase velocity (v_p) and group velocity (v_g) of a travelling wave and show that

$$\frac{1}{v_p} - \frac{1}{v_g} = \frac{\omega}{v_p^2}\frac{dv_p}{d\omega}$$

and that

$$v_g = v_p - \lambda\frac{dv_p}{d\lambda}$$

where λ is the wavelength at frequency $(\omega/2\pi)$.

A dielectric medium has complex relative permittivity $k'_e - j\,k''_e$, where k'_e and k''_e are both frequency dependent; and $k''_e \ll k'_e$. Show that

$$\frac{1}{v_p} \approx \frac{1}{c}(k'_e)^{1/2}$$

$$\frac{1}{v_g} \approx \frac{1}{v_p}\left(1 + \frac{\omega}{2k'_e}\frac{dk'_e}{d\omega}\right)$$

where

$$\frac{1}{c} = \sqrt{\mu_0\epsilon_0}$$

(U.M.)

4.5. A monochromatic plane wave is normally incident on a slab of dielectric as shown in *Figure 4.1.*

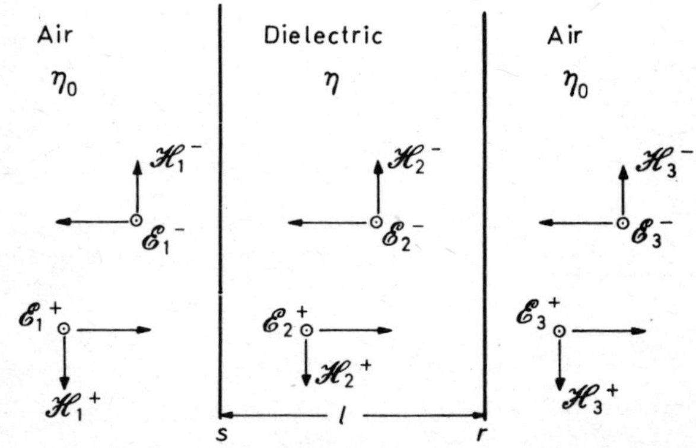

Figure 4.1.

There is air on both sides of the slab and the slab itself may be assumed to be homogeneous, isotropic and characterized by a propagation constant γ and an intrinsic impedance η.

In the steady state, we can consider the total field in each region to be the sum of a forward and backward wave—the electric and magnetic components being designated by $\mathscr{E}^+$, $\mathscr{E}^-$ and $\mathscr{H}^+$, $\mathscr{H}^-$ respectively.

(*a*) Show that the 'wave amplitude matrix', relating these quantities within the slab at planes *s* and *r*, is given by:

$$\begin{bmatrix} \mathscr{E}^+_{s2} \\ \mathscr{E}^-_{s2} \end{bmatrix} = \begin{bmatrix} e^{\gamma l} & 0 \\ 0 & e^{-\gamma l} \end{bmatrix} \begin{bmatrix} \mathscr{E}^+_{r2} \\ \mathscr{E}^-_{r2} \end{bmatrix}; \quad \begin{bmatrix} \mathscr{H}^+_{s2} \\ \mathscr{H}^-_{s2} \end{bmatrix} = \begin{bmatrix} e^{\gamma l} & 0 \\ 0 & e^{-\gamma l} \end{bmatrix} \begin{bmatrix} \mathscr{H}^+_{r2} \\ \mathscr{H}^-_{r2} \end{bmatrix}$$

(*b*) Show also that the 'wave amplitude matrix' relating these quantities across the junction *s* is:

$$\begin{bmatrix} \dfrac{\eta + \eta_0}{2\eta} & \dfrac{\eta - \eta_0}{2\eta} \\ \dfrac{\eta - \eta_0}{2\eta} & \dfrac{\eta + \eta_0}{2\eta} \end{bmatrix}$$

4.6. Apply the results of Problem **4.5** to derive an expression for the reflected wave set up by the slab in *Figure 4.1.* (Assume the dielectric is lossless and that the wave impinging at normal incidence does so from the air onto face *s.*)

4.7. Apply the results of Problem **4.5** to derive an expression for the transmitted wave, $\mathscr{E}^+_{r3}$, in terms of the normally incident wave, $\mathscr{E}^+_{s1}$, for the general case when the slab has losses.

Show that for a slab of thickness $\lambda/2$ measured within the dielectric, having $\epsilon_r = 10$ and $\tan\,\delta = 0{\cdot}02$, the efficiency of energy transmission is about 90%.

4.8. Consider *Figure 4.1* again. If E_s, H_s are the total electric and magnetic field components in the plane *s* and E_r, H_r in the plane *r*, show that these quantities are related by:

$$\begin{bmatrix} E_s \\ H_s \end{bmatrix} = \begin{bmatrix} \cosh \gamma l & \eta \sinh \gamma l \\ \left(\dfrac{1}{\eta}\right) \sinh \gamma l & \cosh \gamma l \end{bmatrix} \begin{bmatrix} E_r \\ H_r \end{bmatrix}$$

Defining the ratio E/H as a 'wave impedance', Z, show that

$$Z_s = \eta\left(\frac{Z_r + \eta \tanh \gamma l}{\eta + Z_r \tanh \gamma l}\right)$$

4.9. Using a Smith Chart, calculate the reflection coefficient from the dielectric slab of *Figure 4.1* in the air region at plane s; $\mathscr{E}_1^+$ is the normally incident wave.

For this problem the slab may be assumed to have the following properties: (*a*) that it is lossless, (*b*) $\epsilon_r = 9$; $\mu = \mu_0$, (*c*) electrical length of slab = $\lambda/12$ (30°), where λ is the wavelength in the slab.

4.10. A dielectric sheet of appropriate thickness can be made to act as a half-silvered mirror (*Figure 4.2*).

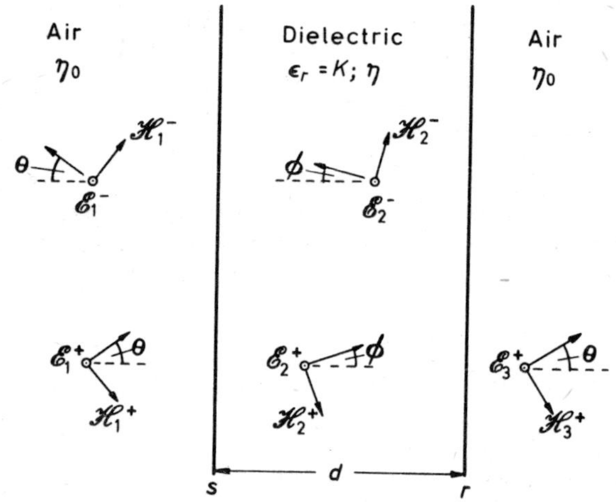

Figure 4.2.

A plane wave of wavelength λ_0 in free space and polarized with its electric vector perpendicular to the plane of incidence strikes the lossless dielectric sheet at an angle of incidence of 45° ($\theta = 45°$).

Show that the reflection coefficient in the air at plane s is given by:

$$\Gamma = -\frac{(K-1)}{K - \mathrm{j}\sqrt{(2K-1)}\cot\left\{\dfrac{2\pi d}{\lambda_0}\sqrt{(K-\tfrac{1}{2})}\right\}}$$

For half power reflection, $|\Gamma| = (1/\sqrt{2})$; hence determine the thickness d of the sheet in the case of ruby mica having dielectric constant $K = 5{\cdot}4$ when $\lambda_0 = 9$mm.

4.11. A uniform plane wave propagating in a lossless dielectric and polarized with its electric vector perpendicular to the plane of incidence strikes the dielectric-air boundary at an angle of incidence θ, θ being greater than the critical angle (*Figure 4.3*).

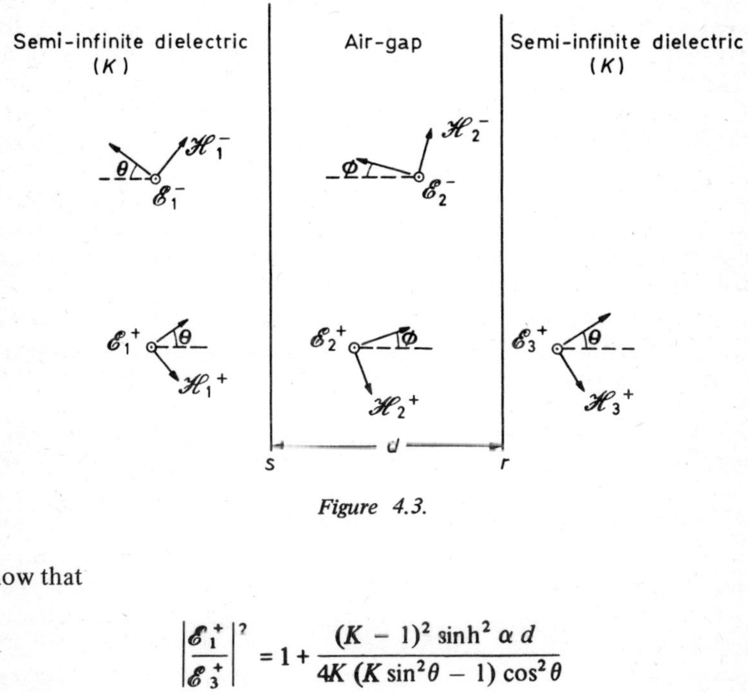

Figure 4.3.

Show that

$$\left|\frac{\mathscr{E}_1^+}{\mathscr{E}_3^+}\right|^2 = 1 + \frac{(K-1)^2 \sinh^2 \alpha d}{4K\,(K \sin^2\theta - 1)\cos^2\theta}$$

where K is the dielectric constant of the dielectric medium and $\alpha = \mathrm{j}\,\beta_0 \cos\phi$.

CHAPTER 5

CIRCUIT CONCEPTS AND QUASI-STATIC APPROXIMATIONS

Introduction

The subject matter of this chapter is essentially halfway between conventional circuit theory and wave theory. In the course with which the authors are concerned some time is devoted to a comprehensive study of the foundations of circuit theory in terms of field theory and an investigation of the limits at which circuit theory may be considered to be invalid. Following this stage methods are developed which enable the conventional theory to be extended in its range of application. The treatment is similar in parts to that of the works listed as References 5 and 6 in which there are some examples of the approximate circuit synthesis approach. These examples have not been repeated here and they should be regarded as supplementary to the problems given here.

The student should observe that there are two forms in which answers to circuit problems may be sought. One of these, as outlined in Reference 6, is to express the behaviour of a component (as a function of frequency) in terms of a more complicated network of components each of which is invariant with frequency. The other, as in some of the problems in this chapter, is to express the component value as a function of frequency. Both representations, when carried out to the same order of approximation are entirely equivalent and the choice of which to use depends on the application.

The other point which should be observed, and which the problems in this chapter should emphasize, is connected with the frequencies at which these methods are required. At first sight it may be thought that only high frequencies are involved. This is by no means the case, however. The experiments on magnetic screening on which there are a few problems are carried out at frequencies less than 20 kHz. When one considers that the skin depth of copper at 50 Hz is only a few millimetres this is perhaps not quite so surprising. The student is recommended to insert some numerical values after solving the general problem so that he gains some idea of the order of magnitude for each of the cases considered.

As far as the solutions to the problems are concerned it will be noticed that two methods have been used. One of these uses the integral laws directly whereas the other uses a power series in frequency from the differential laws. In some of the problems there is little to choose between these two methods and we would suggest that the student becomes familiar with both.

Problems

5.1. Show, using a quasi-static approach, that the first order skin effect correction for the a.c. resistance of a circular wire of radius a is given by:—

$$\frac{R_{\text{a.c.}}}{R_{\text{d.c.}}} = 1 + \frac{1}{48}\left(\frac{a}{\delta}\right)^4$$

where δ is the 'skin depth' factor.

5.2. An engineering problem of some importance is that concerning the power dissipation that occurs in the stator windings of alternating current machines. Suppose we consider for example, an insulated rectangular conductor fitting snugly into an open rectangular slot as shown in *Figure 5.1.*

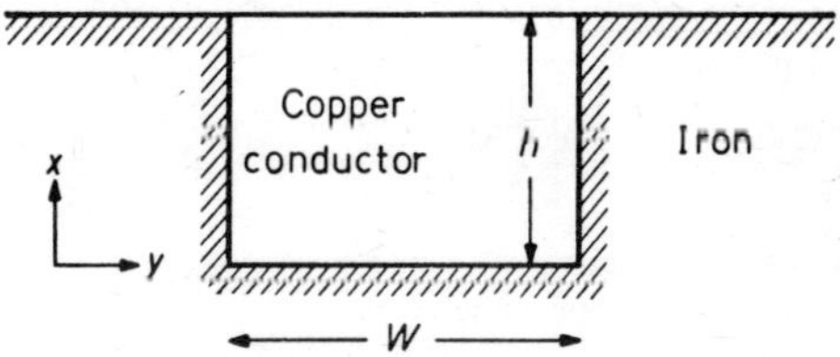

Figure 5.1.

Making any reasonable assumptions, and employing a quasi-static approach, show that the first order skin effect correction for the a.c. resistance of the conductor can be put in the form:

$$\frac{R_{\text{a.c.}}}{R_{\text{d.c.}}} = 1 + \frac{4}{45}s^4$$

where δ is the skin depth factor and $s = (h/\delta)$.

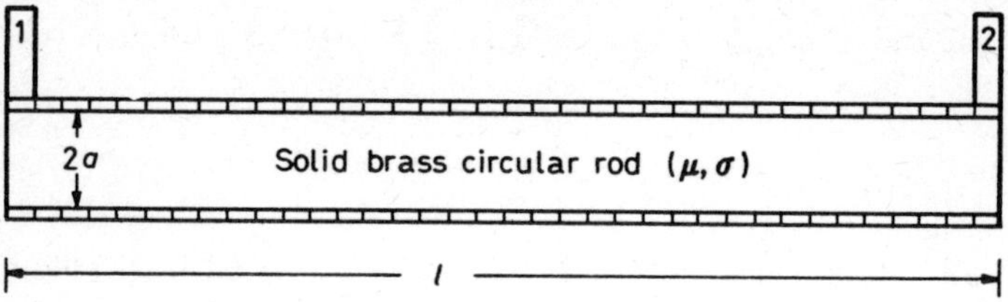

Figure 5.2.

5.3. *Figure 5.2* shows a long coil closely wound onto a solid brass rod of radius a. The coil has N turns and its ohmic resistance may be considered zero.

Using a quasi-static approach, show that the first order impedance of the coil is $j\omega L$ and to the second order it is

$$\omega L\left[j + \frac{1}{4}\frac{a^2}{\delta^2} \right]$$

5.4. An infinitely long, hollow, cylinder of internal and external radii a and b respectively is placed on the axis of an infinite solenoid. Show that at low frequencies, the ratio of the magnitudes of the magnetic field at the outer and inner surfaces of the tube is approximately:

$$\left| \frac{H_{\text{outside}}}{H_{\text{inside}}} \right| \approx 1 + \frac{a^4}{4\delta^4}\left\{ \frac{1}{2}\left(\frac{b^4}{a^4} - 1 \right) - 2 \log_e \frac{b}{a} \right\}$$

where δ is the skin depth factor of the tube.

5.5. Calculate the power loss at low frequencies in the tube mentioned in Problem 5.4.

5.6. The result of Problem 5.4 showed that the magnetic field inside the tube decreases as the frequency increases. If a coil is placed on and perpendicular to the axis it is clear that there will be no induced voltage at zero frequency. Likewise it is clear that, at a high enough frequency, the induced voltage will decrease as the frequency increases due to the screening effect. It is therefore apparent that at some frequency the induced voltage will be a maximum. By calculating the corrections to the fields up to the fourth order in frequency calculate the frequency at which there is a maximum induced voltage.

5.7. Consider a similar system to that of Problem **5.4**, except that there are two coaxial hollow cylinders, each of wall thickness d and of dimensions such that there is an air gap between the cylinders of t. It is stated that the screening effect of this structure is more effective than that of a single cylinder of wall thickness $2d$. Show by means of a flat plate model approximation that this is in fact so and calculate the improvement factor.

5.8. Consider the circular wire of Problem **5.1**. The internal self-inductance of this wire at zero frequency is just $\mu/8\pi$ H/m. Show that, as frequency increases, the first correction to this value obtained is

$$L = \frac{\mu}{8\pi}\left\{1 - \frac{1}{96}\left(\frac{a}{\delta}\right)^4\right\} \text{ H/m}$$

where δ is the skin depth.

CHAPTER 6

FIELDS IN GOOD CONDUCTORS

Introduction

By a good conductor we mean a material in which the displacement current density is negligible in comparison with the conduction current density, i.e. $(\sigma/\omega\epsilon) \gg 1$. Since this definition involves frequency then what is a poor conductor at some frequencies may be a good one at others. When this is so then we can simplify Maxwell's equations by omitting $\partial \mathbf{D}/\partial t$. This chapter can then be thought of as a study of particular cases of Maxwell's equations with this limitation imposed. It leads to some interesting situations which are perhaps not self evident upon a casual inspection.

The equations for **E**, **H** and current density **J** all take the same form, being $\nabla^2 \mathbf{E} = \sigma\mu(\partial \mathbf{E}/\partial t)$. We will only concern ourselves with time variations $e^{j\omega t}$ as usual, although this in itself does not imply any fundamental restrictions to the theory. By writing the solutions of these equations in exponential form we bring out most clearly the wave nature of the solutions and one fact that becomes apparent is that these waves can be very heavily damped. This is not surprising on physical grounds since the Joule heating must abstract energy from the wave in its passage through the material.

A parameter of great practical value is the skin depth factor. This is defined as the depth of penetration for a travelling wave to decrease to e^{-1} of its initial value, which for highly permeable good conductors is only of the order of a mm at power frequencies. Intimately tied up with this is the velocity of propagation of the wave; both phase and group velocity are extremely low. Problem **6.6** illustrates the order of magnitude of the quantities involved.

Another aspect that occurs due to the low intrinsic impedance of a good conductor gives rise to the fact that the energy in a wave travelling in the medium lies almost entirely in the magnetic field and that the magnetic and what little electric component there is are out of phase by 45°.

Many of the problems encountered in practice are at the lower frequencies and are involved in calculating correction factors for skin effect and the like, very much the same sort of problem that is considered in the foregoing chapter. These are not usually phrased to bring out the wave nature of things and they look more like the low frequency circuit problem involving inductances and capacitors.

Problems

6.1. Re-do Problem 5.2 without recourse to quasi-statics.

6.2. Re-do Problem 5.3 without recourse to quasi-statics.

6.3. Define the term 'skin depth' as applied to the flow of electric current.

Show that the ratio of the a.c. resistance R_f to the d.c. resistance R_0 of a circular conductor of radius a is given under certain conditions by

$$\frac{R_f}{R_0} = \frac{a}{2\delta}$$

where δ is the skin depth. State the conditions.

In order to reduce the losses in this conductor it is proposed to reduce the resistivity by cooling. By how much will the losses be reduced if the resistivity is reduced by 99%? (U.Shn.)

6.4. Show that the magnetic field strength in a flat sheet of large surface area is governed by the equation

$$\frac{\partial^2 \mathbf{H}}{\partial y^2} = \frac{\mu_0 \mu_r}{\rho} \frac{\partial \mathbf{H}}{\partial t}$$

if the applied field is uniform and parallel to the surface.

Show that the eddy current in such a sheet will reduce the alternating magnetic field at the centre so that the ratio of the magnitudes of surface and central field is given by the equation

$$\left|\frac{H_s}{H_0}\right|^2 = \frac{1}{2}\left\{\cosh\frac{2a}{\delta} + \cos\frac{2a}{\delta}\right\}$$

where $2a$ is the thickness of the sheet, $\delta = \sqrt{(2\rho/\mu_0\mu_r\omega)}$ and ω is the angular frequency of the magnetic field. (U.Shn.)

6.5. A copper conductor of strip form has a length and breadth which are both much greater than the thickness $2b$. A coordinate system is taken with the origin at the centre of all three dimensions of the strip, the axes of x, y and z being in the directions of the breadth, the thickness and the length respectively. It can be assumed that under a.c. conditions the current density and magnetic field phasors obey a differential equation of the form

$$\frac{d^2J}{dy^2} - \frac{j}{d^2} J = 0,$$

Prove that, when the current density phasor J is entirely in the z direction, its value at any point is given by

$$J = J_0 \cosh \frac{(1 + j)y}{d\sqrt{2}},$$

where J_0 is the (r.m.s.) value at the centre. Neglecting edge effects, prove also that

$$H = H_x = - J_0 \, d \left(\frac{1 - j}{\sqrt{2}} \right) \sinh \frac{(1 + j)y}{d\sqrt{2}}$$

Hence, by applying Poynting's method, prove that the power consumed per unit length and breadth is

$$\rho \, J_0{}^2 \, d \sqrt{2} (\sinh \theta \cosh \theta + \sin \theta \cos \theta)$$

where

$$\theta = \frac{b}{d\sqrt{2}}$$

$$[\cosh (1 + j)\theta \equiv \cosh \theta \cos \theta + j \sinh \theta \sin \theta;$$

$$\sinh (1 + j)\theta \equiv \sinh \theta \cos \theta + j \cosh\theta \sin \theta.]$$ (U.Ld.)

6.6. (*a*) Starting from Maxwell's equations derive expressions for the phase and attenuation constants, phase and group velocities, skin depth factor and intrinsic impedance for a plane wave propagating in a 'good conductor'.

(*b*) Typical values for the conductivity and permeability of good quality transformer steel may be taken as 2×10^6 $1/\Omega$ m and 2,500 μ_0 respectively. Calculate the above constants at 50 Hz.

6.7. A semi-infinite block of conducting material, region (2), occupies the half-plane to the right of the yz plane of an xyz coordinate system (*Figure*

6.1). The surface is coated to a depth δ with a material of better conductivity.

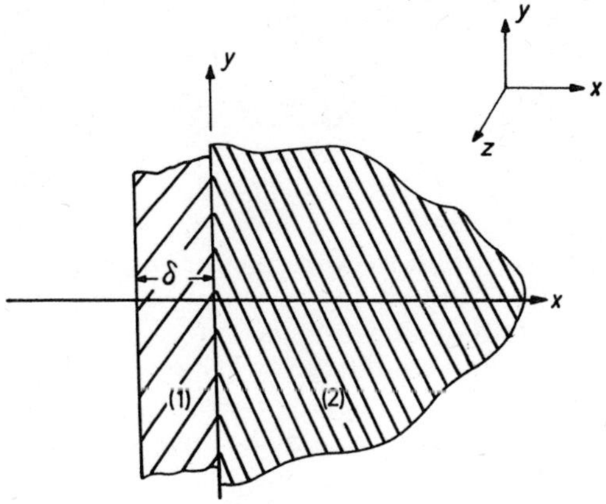

Figure 6.1.

High frequency alternating current flows in the z direction $[J_z = f(x, t)]$ such that the current density J_z just in region (2) is J_0. Deduce expressions for the variations of J_z in the two regions and sketch the form of this variation. (U.A.)

CHAPTER 7

WAVEGUIDES AND RESONANT STRUCTURES

Introduction

There is very little which needs to be said in introduction to the problems in this chapter. The subject matter is of sufficient practical importance, aside from academic interest, to speak for itself.

One of the main problems the student will meet in his study of this topic is the apparently loose way many authors use the names and the symbols for the fields. For instance, E may be used in the same book for:

(*a*) the physical electric field,

(*b*) the phasor vector field (that is, the complex vector with the time factor suppressed),

(*c*) what may be called the reduced phasor vector field (the propagation factor $\exp(-jkz)$ may also be suppressed).

The term 'electric field' will also be used for all of these quantities.

The student should understand at this stage that the process is not as loose as it appears. Fundamentally these authors are talking about physical quantities and the form in which they express them is understood from the context. All the forms are physically equivalent and the student may be assured that, with familiarity, the difficulty will vanish.

In the selection of problems we have avoided those which could be classed as microwave technique (except possibly for Problem 7.9). This type of problem can be found in an excellent modern book on the subject (Reference 7) and we would suggest that the student solves some of these in addition to the problems of this chapter.

Problems

7.1. A guiding structure consists of two or more perfect conductors, each of arbitrary cross-section that is uniform in the z-direction. Prove directly from Maxwell's equations that for TEM waves propagating along this structure, the electric and magnetic fields can be derived from the corresponding electrostatic and magnetostatic solutions.

The electric and magnetic fields can exist quite independently if they do not vary with time. Show how they are completely dependent when varying with angular frequency ω.

This last statement suggests that in an a.c. mains supply cable, the voltage and current must have a definite ratio. There is no such restriction in practice. Explain the paradox. (U.S.)

7.2. A wave propagates in the positive z-direction with given frequency and propagation constant. Prove directly from Maxwell's equations that if the wave has a known E_z component, with $H_z = 0$, the whole electromagnetic field can be derived.

Such a wave is given by

$$E_z = E_0 \sin \frac{\pi x}{x_0} \sin \frac{\pi y}{y_0} \exp \mathrm{j}(\omega t - \beta z)$$

Derive the remaining field components.

Where could conducting walls be placed so as to contain the above fields, as in a waveguide?

Give the cross-section dimensions of the smallest of such waveguides, and calculate the associated cut-off frequency. (U.S.)

7.3. The fields of the dominant TE_{01} mode in rectangular waveguide are of the form

$$E_x = E_0 \cos (ky \sin \theta) \exp \mathrm{j}(\omega t - kz \cos \theta)$$

$$H_y = H_0 \cos \theta \cos (ky \sin \theta) \exp \mathrm{j}(\omega t - kz \cos \theta)$$

$$H_z = H_0 \,\mathrm{j} \sin \theta \sin (ky \sin \theta) \exp \mathrm{j}(\omega t - kz \cos \theta)$$

where $k^2 = \omega^2 \mu\epsilon$ and $E_0/H_0 = \sqrt{(\mu/\epsilon)}$.

If the waveguide walls are at $x = \pm a$ and $y = \pm b$, what is the cut-off wavelength in terms of these guide dimensions.

From the above field expressions, derive the relation between free-space, cut-off and guide wavelengths.

From the above expressions, describe whether or not the waves would be disturbed by narrow slits cut in the waveguide walls $y = \pm b$, with the slits parallel to (a) the x-axis, (b) the z-axis.

Calculate the total time-average power flow along the waveguide. (U.S.)

7.4. A circular waveguide has a radius a. It carries a TE mode of propagation in which the field vectors are independent of the angular coordinate. Derive the fields for the lowest mode of this type and calculate the attenuation of the guide for this mode.

7.5. Consider the circular waveguide mode of the previous problem. Why is the corresponding mode of this type suitable for use in a tunable resonant cavity and how could a transmission cavity be coupled at the input and output to eliminate the effect of other modes? Calculate the fields of the first resonant mode of such a cavity and determine the unloaded Q.

7.6. A coaxial line has an inner conductor of radius a and an outer conductor of radius b. Derive expressions for the electric and magnetic fields in this line for the TEM mode of propagation. Calculate also the attenuation coefficient and the characteristic impedance of the line.

If the line is to be used in an application which necessitates a maximum outer conductor radius b_0 calculate the inner conductor radius and characteristic impedance of (*a*) the transmission line with the lowest attenuation and (*b*) the line which will withstand the greatest power assuming that there is no reflected wave.

7.7. A rectangular waveguide is filled with a dielectric material of dielectric constant ϵ_r, except for a section of length s which is empty, as shown in *Figure 7.1*. The dimensions of the guide are chosen such that the dominant

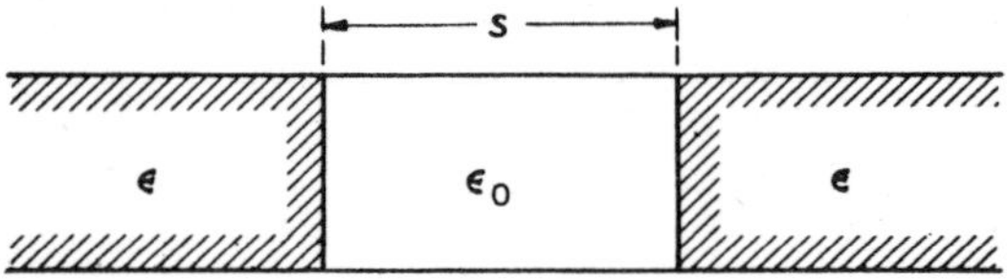

Figure 7.1.

TE_{10} mode propagates in the dielectric filled region but is evanescent in the empty region. A wave is incident from the left and a matched load terminates the guide at the right.

Calculate the power dissipated in the load (in relation to the power in the incident wave) as a function of the dimension s.

7.8. The fields of the TE_{10} mode in a rectangular waveguide with broad dimension a and narrow dimension b are given by

$$E_y = A \sin \frac{\pi x}{a} \exp\{-j\beta z\}$$

$$H_x = -\frac{\beta}{\omega\mu} A \sin \frac{\pi x}{a} \exp\{-j\beta z\}$$

$$H_z = +j\frac{(\pi/a)}{\omega\mu} A \cos \frac{\pi x}{a} \exp\{-j\beta z\}$$

The characteristic impedance of a transmission line can usually be determined from a calculation of

(*a*) power and longitudinal current
or (*b*) power and transverse voltage
or (*c*) transverse voltage and longitudinal current

and for TEM transmission lines these three alternatives lead to the same answer.

Show that for the TE_{10} waveguide three different answers are obtained. (The transverse voltage is usually defined as the line integral of the electric field at the centre of the broad face of the waveguide.)

7.9. Two rectangular waveguides of the same broad dimension but of different narrow dimensions b and b' are connected together. The equivalent circuit of such a structure can be shown to be that of *Figure 7.2*. What relationship would you expect to find between Y_0 and Y_0'?

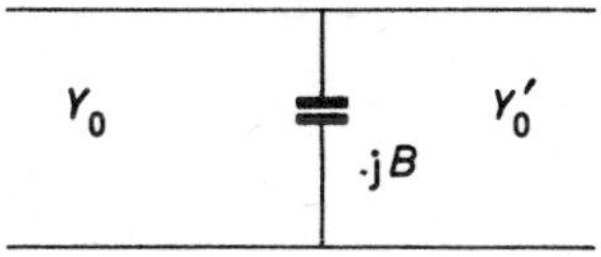

Figure 7.2.

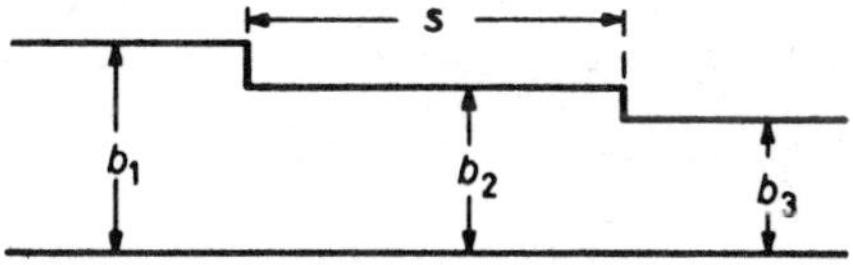

Figure 7.3.

Consider the structure of *Figure 7.3* in which three rectangular waveguides are connected, having common broad dimensions. Assuming that the shunt susceptances at the junctions (i.e. those corresponding to jB of *Figure 7.2*) are small enough to be ignored and that the dimensions b_1 and b_3 are pre-determined, what dimensions should b_2 and s take so that the composite junction is reflectionless at a given frequency?

7.10. A rectangular conducting plate of linear dimensions a and b and of negligible thickness is placed between two conducting ground planes spaced apart by a distance d. The plate is centrally disposed with respect to the ground planes and is excited by an electromagnetic field which is symmetrical about the plane of the plate. If a, b, $\gg d$ so that the fringing fields may be neglected, calculate the frequencies at which the plate will resonate. (It may be assumed that $d \ll \lambda_0$.) How would you expect the answer to be modified by the presence of the fringing fields?

CHAPTER 8

FIELDS AND MOVING MEDIA

Introduction

The subject matter of this chapter will require the student to think very clearly in order to be correct about the various fields which are observed in stationary (relative to the observer) and moving coordinate systems. A very good treatment of the subject will be found in some of the early chapters of Reference 4 although the book also deals with the second order effects of relativity which are outside the scope of the problems following. Essentially the problems of this chapter all assume that terms of order $(v/c)^2$ are negligible. This fact leads to inconsistency in thc equations used to transform the fields. It is worthwhile at this stage reviewing these equations. If we denote by **E**, **B**, **D** and **H** the electromagnetic field quantities measured by an observer in a given system of coordinates Σ and by $\mathbf{E}'$, $\mathbf{B}'$, $\mathbf{D}'$ and $\mathbf{H}'$ the corresponding quantities measured by an observer in another system Σ' which moves relative to Σ with a uniform velocity **v**, then the equations relating the various quantities are

$$\mathbf{E}' = \mathbf{E} + \mathbf{v} \times \mathbf{B} \qquad \mathbf{B}' = \mathbf{B} - \frac{\mathbf{v} \times \mathbf{E}}{c^2}$$

$$\mathbf{D}' = \mathbf{D} + \frac{\mathbf{v} \times \mathbf{H}}{c^2} \qquad \mathbf{H}' = \mathbf{H} - \mathbf{v} \times \mathbf{D}$$

Now since Σ is moving relative to Σ' with a velocity of $-\mathbf{v}$ we can immediately write the inverse equations as

$$\mathbf{E} = \mathbf{E}' - \mathbf{v} \times \mathbf{B}' \qquad \mathbf{B} = \mathbf{B}' + \frac{\mathbf{v} \times \mathbf{E}'}{c^2}$$

$$\mathbf{D} = \mathbf{D}' - \frac{\mathbf{v} \times \mathbf{H}'}{c^2} \qquad \mathbf{H} = \mathbf{H}' + \mathbf{v} \times \mathbf{D}'$$

These equations can be substituted back into the first set and we shall find almost immediately that the equations are not satisfied. The errors are, however, of order $(v/c)^2$ and within our approximation noted above would then be considered valid. (Needless to say, the equations corrected for relativistic effects are consistent.)

Some of the problems are chosen from examination papers and others have been selected which, while possibly being classed as bookwork, do illuminate some of the more important concepts. Two of the problems deal with a possible practical instrument. The student could, with benefit, extend these to include the use of conducting liquids and other electromagnetic arrangements.

One word of advice to the student: the secret of being able to solve these problems lies almost entirely in the careful thought devoted to the transformation of the fields from an observer in one system to an observer in another. Remember that the physical system itself is independent of the coordinate system but the method by which we describe it is not.

Problems

8.1. Show that a moving magnetic dielectric may exhibit a polarization, part of which is true and part apparent.

A flat plate of magnetic dielectric is situated in a transverse electric field and a tangential magnetic field and moves with a velocity $\mathbf{v} \ll c$ perpendicularly to both of these fields. If the field components due to external sources are $\mathbf{E}_0$ and $\mathbf{B}_0$, show that the moving plate exhibits a polarization given by

$$\mathbf{P} = \epsilon_0 \left\{ \left(1 - \frac{1}{K}\right) \mathbf{E}_0' + \left(\mu - \frac{1}{K}\right) \mathbf{v} \times \mathbf{B}_0 \right\}$$

where K and μ are the permittivity and permeability of the plate. (U.St.A.)

8.2. A parallel plate capacitor is constructed from two circular plates separated by a dielectric disc of permittivity ϵ. The whole assembly is constructed in such a fashion that

(*a*) the plates can be rotated with the dielectric remaining stationary;

(*b*) the dielectric can be rotated with the plates stationary; and

(*c*) both the plates and the dielectric can be rotated with the same angular velocity.

The capacitor is charged to a potential V. Investigate the magnetic fields produced in the three cases above and relate them to the magnetic field intensity $\mathbf{H}$ and the magnetisation $\mathbf{M}$. (Any gap between the dielectric and the plates can be ignored.)

8.3. A fuel pipe carrying ethyl alcohol (dielectric constant of 26) is built from a thin-walled insulating tube of square cross section. In contact with two opposite walls of the tube are the poles of a magnet which produces a flux density of 0·1 Wb/m^2 across the gap of 2 cm. In the other two walls there are two conducting electrodes which are connected to a voltmeter of very high input impedance. If the alcohol is flowing down the tube with a velocity of 5 m/s, calculate the voltage developed across the electrodes. (Assume for simplicity that the walls of the tube perpendicular to the electrodes are magnetic walls for the field problem.)

8.4. In the previous problem, if the magnet is removed and a coil of 100 turns with a cross-sectional area of 1 cm^2 is placed against one of the sides, and if the voltmeter is replaced by a sinusoidal voltage source of 1000 V peak at 10 MHz, calculate the voltage induced in the coil.

8.5. Show that the differential form of Faraday's Law when applied to a system moving with a velocity **u** can be written

$$\nabla \times \mathbf{E} = -\frac{\partial \mathbf{B}}{\partial t} + \nabla \times \mathbf{u} \times \mathbf{B}$$

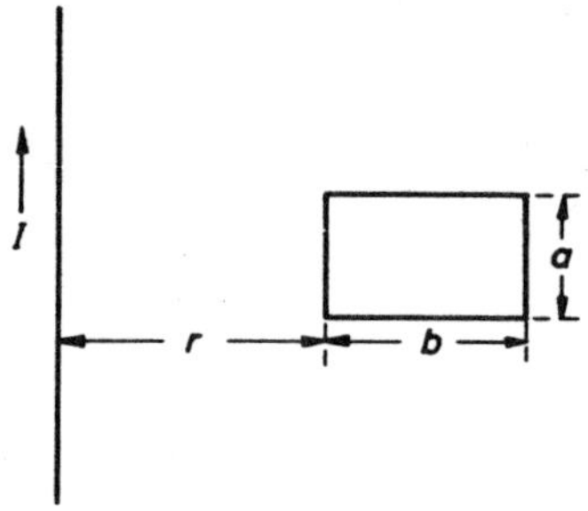

Figure 8.1.

A rectangular loop is located in the field of a long current-carrying straight wire as shown (*Figure 8.1*). Use the above expression to find the following:

(i) The induced voltage in the loop if it is fixed in space but I varies as $I_0 \cos(\omega t)$.

(ii) The magnitude and direction of the induced voltage as a function of r when $I = I_0$ and is constant but the loop moves towards the wire with velocity **u**.

(iii) The induced voltage in the loop when it moves towards the long current conductor with velocity **u** and the current varies as $I = I_0 \cos(\omega t)$.

(U.S.)

8.6. A conducting permanent magnet, magnetized axially and axially symmetrical, rotates about the axis with uniform angular velocity ω. Show that the electric field intensity at a point inside the magnet is given by

$$\mathbf{E} = -\text{grad}\,(\mathbf{v}\,.\,\mathbf{A})$$

where

$$\mathbf{v} = \boldsymbol{\omega} \times \mathbf{r}$$

and

$$\text{curl}\,\mathbf{A} = \mathbf{B}$$ (U.St.A.)

8.7. Consider a moving body, moving with velocity **v**, and a vector field **A** which is fixed *in the body* but varies with position in the body. If $\partial\mathbf{A}/\partial t$ measures the time rate of change of **A** at a point fixed in the *stationary* reference frame, show that

$$\frac{\partial \mathbf{A}}{\partial t} = -(\mathbf{v}\,.\,\nabla)\mathbf{A}$$

8.8. A single loop of fine wire carries a constant current I and causes a magnetic field of flux density **B** at any given point. When the loop moves with constant velocity $v \ll c$ in a straight line, show that, if the motion produces an electric field given by $\mathbf{E} = \mathbf{B} \times \mathbf{v}$, the loop must carry a linear charge whose value, per unit length of conductor, is given by

$$\sigma = \frac{\mathbf{v}\,.\,\mathbf{I}}{c^2}$$

Verify that this apparent charge accounts for the whole of the electric field produced by the longitudinal motion of a long transmission line which consists of two parallel flat conducting plates, each of width a and separated by a distance $b \ll a$. The line carries a steady current I. (U.St.A.)

PART II

SOLUTIONS

CHAPTER 1

1.1. Consider for example

$$\mathbf{A}(\mathbf{r},t) = \frac{\mu_0}{4\pi}\int_v \frac{[\mathbf{J}]}{|\mathbf{r}-\mathbf{r}'|}\,\mathrm{d}V$$

which is an expression for the magnetic vector potential due to a current distribution of density **J** distributed throughout some volume V.

What this expression means physically is that the contribution to the potential at some field point **r** at time t, due to a volume current element at some source point $\mathbf{r}'$, is due not to the current at $\mathbf{r}'$ at the same instant of time t but to the current density at the earlier time $t - |\mathbf{r}-\mathbf{r}'|/c$ . . . . this is the reason attributable to the squared brackets; hence the name 'retarded potential'.

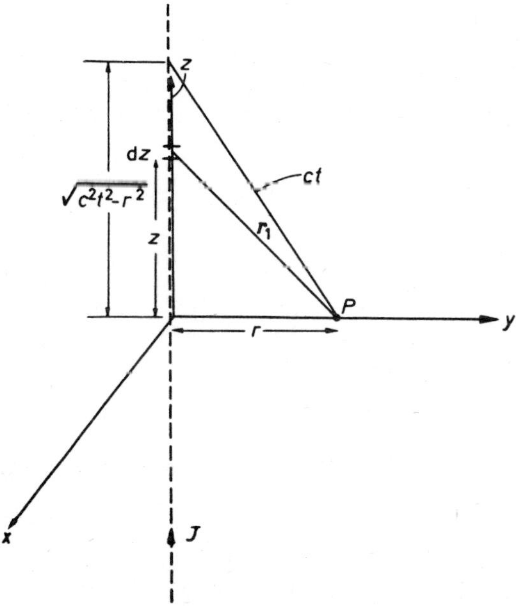

Figure 1.2.

Figure 1.2 shows the geometry involved, where such a current distribution is linear, the volume integration is reduced to a simple integral.

If we now think of the electromagnetic disturbances as propagating from the current element sources with a finite velocity, c, then it follows that at the instant of time t after switching on the current the farthest distance from the field point which a source can contribute to the field is simply ct. Hence

$$\mathbf{A} = \mathbf{k}\frac{\mu_0 J}{2\pi}\int_0^{\sqrt{(c^2t^2-r^2)}} \frac{dz}{\sqrt{(r^2+z^2)}}$$

$$= \mathbf{k}\frac{\mu_0 J}{2\pi}\log_e\{\sqrt{(z^2+r^2)}+z\}\Big|_0^{\sqrt{(c^2t^2-r^2)}}$$

$$\therefore \qquad \mathbf{A} = \mathbf{k}\frac{\mu_0 J}{2\pi}\log_e\left\{\frac{\sqrt{(c^2t^2-r^2)}+ct}{r}\right\}, \qquad t \geqslant \frac{r}{c}$$

and

$$\mathbf{A} = 0, \quad t < \frac{r}{c}$$

The magnetic field at $\mathbf{r}$ is readily found by taking the curl of $\mathbf{A}$, which in cylindrical coordinates and for the single $\mathbf{k}$ component of $\mathbf{A}$ reduces to

$$\mathbf{B} = -\frac{\partial A}{\partial r}\hat{\boldsymbol{\phi}}$$

$$= -\hat{\boldsymbol{\phi}}\frac{\mu_0 J}{2\pi}\left\{\frac{\frac{1}{2}(c^2t^2-r^2)^{-1/2}(-2r)}{ct+\sqrt{(c^2t^2-r^2)}} - \frac{1}{r}\right\}$$

$$\therefore \qquad \mathbf{B} = \hat{\boldsymbol{\phi}}\frac{\mu_0 ctJ}{2\pi r\sqrt{(c^2t^2-r^2)}} \qquad . \, t \geqslant \frac{r}{c}$$

Since the current flow is steady and there are no free charges on the wire, we have

$$\mathbf{E} = -\frac{\partial \mathbf{A}}{\partial t} = -\frac{\mu_0 J}{2\pi}\left\{\frac{c+\frac{1}{2}(c^2t^2-r^2)^{-1/2}2c^2t}{ct+\sqrt{(c^2t^2-r^2)}}\right\}\mathbf{k}$$

$$\therefore \qquad \mathbf{E} = -\frac{\mu_0 cJ\mathbf{k}}{2\pi\sqrt{(c^2t^2-r^2)}} \qquad . \, t \geqslant \frac{r}{c}$$

We note as a check in passing that these expressions do simplify to the static solutions as $t \to \infty$, i.e.

$$\mathbf{E} \to 0 \quad \text{and} \quad \mathbf{B} \to \frac{\mu_0 J}{2\pi r}\hat{\boldsymbol{\phi}}$$

To determine the flux of energy across unit length of a cylinder through *P* we employ Poynting's Theorem.

$$\mathbf{S} = \mathbf{E} \times \mathbf{H} = -\frac{\mu_0 cJ\mathbf{k}}{2\pi\sqrt{(c^2t^2 - r^2)}} \times \frac{Jct\,\hat{\boldsymbol{\phi}}}{2\pi r\sqrt{(c^2t^2 - r^2)}}$$

$$\therefore \text{Energy flux (in } \mathbf{watts}) = \frac{\mu_0 J^2 c^2 t}{2\pi(c^2t^2 - r^2)}$$

1.2. The retarded potential solution of the inhomogeneous equation can be found in the reference cited and will not be repeated here. For our problem, we have

$$\mathbf{A} = \frac{\mu_0}{4\pi}\oint_c \frac{[\mathscr{I}]}{|\mathbf{r}|}\,\mathrm{d}\mathbf{l}$$

where c is the circle of radius a (see *Figure 1.3*).

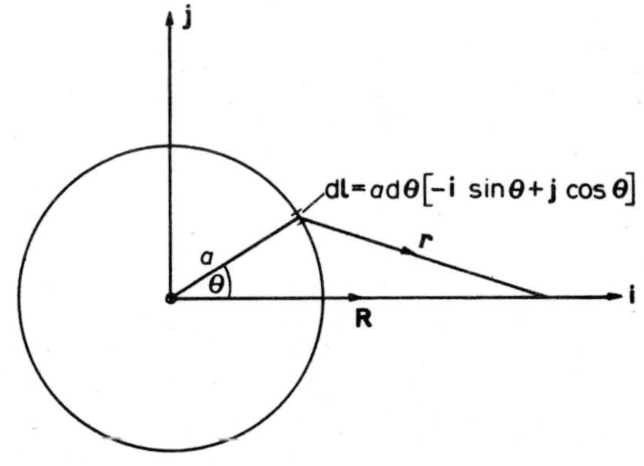

Figure 1.3.

Consider the current to be of the form $\mathscr{I} = Ie^{j\omega t}$ and $\mathbf{dl} = a d\theta\ [-\mathbf{i}\sin\theta + \mathbf{j}\cos\theta]$. Then, the component of $\mathbf{A}$ perpendicular to the radius vector can be found by calculating A_j at $\theta = 0$, i.e.

$$A_j = \frac{\mu_0 a I e^{j\omega t}}{4\pi}\int_0^{2\pi}\frac{e^{-j(\omega/c)\sqrt{(R^2 - 2aR\cos\theta + a^2)}}\cos\theta\, d\theta}{\sqrt{(R^2 - 2aR\cos\theta + a^2)}}$$

$$\doteqdot \frac{\mu_0 a I e^{j\omega t}}{4\pi R}\int_0^{2\pi}\frac{e^{-j(\omega R/c)\sqrt{[1 - (2a/R)\cos\theta]}}\cos\theta\, d\theta}{\sqrt{[1 - (2a/R)\cos\theta]}}$$

where we have ignored terms in $(a/R)^2$.

Using a Binomial expansion and retaining only the first order terms, the integral can be approximated by

$$A_j \doteqdot \frac{\mu_0 a I e^{j\omega[t - (R/c)]}}{4\pi R}\int_0^{2\pi}\left(1 + j\frac{\omega a}{c}\cos\theta\right)\left(1 + \frac{a}{R}\cos\theta\right)\cos\theta\, d\theta$$

$$= \frac{\mu_0 a I e^{j\omega[t - (R/c)]}}{4\pi R}\int_0^{2\pi}\left[\cos\theta + \left(\frac{a}{R} + j\frac{\omega a}{c}\right)\cos^2\theta + j\frac{\omega a^2}{cR}\cos^3\theta\right] d\theta$$

$$= \frac{\mu_0 a I e^{j\omega[t - (R/c)]}}{4R}\cdot\left[\frac{a}{R} + j\frac{\omega a}{c}\right]$$

Thus for a current of the form,

$$\mathscr{I} = I\cos\omega t$$

then

$$A_j = \frac{\mu_0 I a^2}{4R}\left\{\frac{1}{R}\cos\omega\left(t - \frac{R}{c}\right) - \frac{\omega}{c}\sin\omega\left(t - \frac{R}{c}\right)\right\}$$

1.3. Maxwell's equations in free space are

$$\nabla \times \mathbf{E} = -\frac{\partial \mathbf{B}}{\partial t} \qquad \ldots.(1)$$

$$\nabla \times \mathbf{H} = \epsilon_0\frac{\partial \mathbf{E}}{\partial t} \qquad \ldots.(2)$$

$$\nabla \cdot \mathbf{E} = 0 \qquad \ldots.(3)$$

$$\nabla \cdot \mathbf{B} = 0 \qquad \ldots.(4)$$

Since $\nabla . \mathbf{B} = 0$, we can put

$$\mathbf{B} = \nabla \times \mathbf{A} \qquad \ldots . (5)$$

Substituting Equation (5) into (1) gives

$$\nabla \times \left(\mathbf{E} + \frac{\partial \mathbf{A}}{\partial t} \right) = 0$$

$$\therefore \mathbf{E} + \frac{\partial \mathbf{A}}{\partial t} = -\nabla \phi$$

giving

$$\mathbf{E} = -\nabla \phi - \frac{\partial \mathbf{A}}{\partial t} \qquad \ldots . (6)$$

Equations (5) and (6) are well known expressions for the field vectors in terms of the electric scalar potential ϕ and the magnetic vector potential **A**.

Using Equations (2), (5) and (6) together with the identity $\nabla \times \nabla \times \mathbf{A} = \nabla(\nabla . \mathbf{A}) - \nabla^2 \mathbf{A}$, we obtain

$$\nabla(\nabla . \mathbf{A}) - \nabla^2 \mathbf{A} + \frac{1}{c^2} \nabla \left(\frac{\partial \phi}{\partial t} \right) + \frac{1}{c^2} \frac{\partial^2 \mathbf{A}}{\partial t^2} = 0 \qquad \ldots . (7)$$

Similarly, from Equations (3) and (6)

$$\nabla^2 \phi + \frac{\partial}{\partial t} (\nabla . \mathbf{A}) = 0 \qquad \ldots . (8)$$

Equations (7) and (8) are mixed, i.e. they both contain **A** and ϕ and we would like to separate them. We can do just this if we note that **A** and ϕ as defined are not unique and therefore impose a further condition (the Lorentz condition) that

$$\nabla . \mathbf{A} = -\frac{1}{c^2} \frac{\partial \phi}{\partial t}$$

The separated equations are

$$\nabla^2\phi = \frac{1}{c^2}\frac{\partial^2\phi}{\partial t^2}$$

and
$$\nabla^2\mathbf{A} = \frac{1}{c^2}\frac{\partial^2\mathbf{A}}{\partial t^2}$$

Comparing the equations given with the above,

i.e.
$$\mathbf{E} = \nabla\left(\frac{\partial V}{\partial z}\right) - \mathbf{k}\frac{1}{c^2}\frac{\partial^2 V}{\partial t^2}$$

where

$$\nabla^2 V = \frac{1}{c^2}\frac{\partial^2 V}{\partial t^2}$$

we see that they do lead to a possible electromagnetic field. In this particular case,

$$\phi = -\frac{\partial V}{\partial z}$$

$$\mathbf{A} = \frac{1}{c^2}\frac{\partial V}{\partial t}\mathbf{k}$$

and
$$\mathbf{H} = \frac{1}{\mu_0}(\nabla \times \mathbf{A})$$

and we can see by inspection that all the conditions are satisfied.

If
$$V = \frac{1}{r}\sin\frac{r}{a}\sin\frac{ct}{a}$$

then
$$\frac{1}{c^2}\frac{\partial^2 V}{\partial t^2} = -\frac{1}{a^2}\frac{1}{r}\sin\frac{r}{a}\sin\frac{ct}{a} = -\frac{V}{a^2}$$

and
$$\nabla^2 V = \frac{1}{r}\left[-\frac{1}{a^2}\sin\frac{r}{a}\sin\frac{ct}{a}\right] = -\frac{V}{a^2}$$

therefore the equation for V is satisfied.

Now $$\mathbf{E} = \nabla\left(\frac{\partial V}{\partial z}\right) - \mathbf{k}\frac{1}{c^2}\frac{\partial^2 V}{\partial t^2} = \frac{\partial}{\partial z}(\nabla V) + \mathbf{k}\frac{V}{a^2}$$

and $$\frac{\partial}{\partial z}(\nabla V) = \frac{\partial}{\partial z}\left\{\frac{1}{r}\frac{\partial V}{\partial r}\mathbf{r}\right\} = \frac{\partial}{\partial z}\left\{\left(\frac{1}{r}\frac{\partial V}{\partial r}\right)(\mathbf{i}x + \mathbf{j}y + \mathbf{k}z)\right\}$$

$$= \frac{z}{r}\frac{\partial}{\partial r}\left\{\frac{1}{r}\frac{\partial V}{\partial r}\right\}\mathbf{r} + \frac{1}{r}\frac{\partial V}{\partial r}\mathbf{k}$$

$\therefore$ $$\mathbf{E} = \mathbf{k}\left\{\frac{1}{r}\frac{\partial V}{\partial r} + \frac{V}{a^2}\right\} + \frac{z}{r}\frac{\partial}{\partial r}\left\{\frac{1}{r}\frac{\partial V}{\partial r}\right\}\mathbf{r}$$

1.4. The first part of this problem has already been covered in the previous solution.

Now $\mathbf{A} = \mathbf{k}\ [f'(u)/cr]$ must satisfy the condition

$$\nabla^2\left\{\frac{f'(u)}{cr}\right\} = \frac{1}{c^2}\frac{\partial^2}{\partial t^2}\left\{\frac{f'(u)}{cr}\right\}$$

In spherical coordinates, and making use of symmetry,

$$\nabla^2\left\{\frac{f'(u)}{cr}\right\} = \frac{1}{r^2}\frac{\partial}{\partial r}\left[r^2\frac{\partial}{\partial r}\left\{\frac{f'(u)}{cr}\right\}\right]$$

$\therefore$ $$r^2\frac{\partial}{\partial r}\left\{\frac{f'(u)}{cr}\right\} = \frac{r^2}{c}\left\{\frac{1}{r}\frac{\partial}{\partial r}[f'(u)] + f'(u)\left(-\frac{1}{r^2}\right)\right\}$$

$$= -\frac{1}{c}\left\{f'(u) + \frac{r}{c}f''(u)\right\}$$

$\therefore$ $$\frac{1}{r^2}\frac{\partial}{\partial r}\left\{r^2\frac{\partial}{\partial r}\left[\frac{f'(u)}{cr}\right]\right\} = \frac{1}{c^3}\frac{1}{r}f'''(u)$$

also $$\frac{1}{c^2}\frac{\partial^2}{\partial t^2}\left\{\frac{1}{cr}f'(u)\right\} = \frac{1}{c^3 r}f'''(u)$$

Hence we see that the necessary condition is satisfied.

To determine a form for the scalar potential we can use the Lorentz condition connecting **A** with ϕ.
i.e.

$$\nabla . \mathbf{A} = \frac{1}{-c^2}\frac{\partial \phi}{\partial t}$$

$$\frac{\partial}{\partial z}\left\{\frac{f'(u)}{cr}\right\} = \frac{1}{-c^2}\frac{\partial \phi}{\partial t}$$

thus a possible form is given by

$$\phi = -c\frac{\partial}{\partial z}\left[\frac{f(u)}{r}\right]$$

1.5. The first part of this problem involves obtaining the boundary conditions that a field must satisfy at an air-metal surface and since the reader is assumed to know these already, a proof of these need not be repeated here. If in doubt, a proof can be found in works given in the References.

Let us consider whether the given field expressions satisfy Maxwell's equations, which for free space reduce to

$$\nabla \times \mathbf{E} = -\mu_0 \frac{\partial \mathbf{H}}{\partial t} \quad \ldots .(1)$$

$$\nabla \times \mathbf{H} = \epsilon_0 \frac{\partial \mathbf{E}}{\partial t} \quad \ldots .(2)$$

$$\nabla . \mathbf{E} = 0 \quad \ldots .(3)$$

$$\nabla . \mathbf{B} = 0 \quad \ldots .(4)$$

We can see by inspection that (3) and (4) are satisfied.

Now

$$\nabla \times \mathbf{E} = -\mathbf{j}\frac{\partial E_z}{\partial x} = -\mathbf{j}An \cos n(x - ct)$$

and

$$-\mu_0 \frac{\partial \mathbf{H}}{\partial t} = -\mathbf{j}\mu_0 \lambda Anc \cos n(x - ct)$$

Therefore Equation (1) is satisfied if $\mu_0 c\lambda = 1$ $\ldots .(5)$

Also $$\nabla \times \mathbf{H} = \mathbf{k}\frac{\partial H_y}{\partial x} = -\mathbf{k}\lambda An \cos n(x - ct)$$

and $$\epsilon_0 \frac{\partial \mathbf{E}}{\partial t} = -\mathbf{k}\, \epsilon_0 cnA \cos n(x - ct)$$

Therefore Equation (2) is satisfied if $\epsilon_0 c = \lambda$ .... (6)
From Equations (5) and (6)

$$c = \frac{\lambda}{\epsilon_0} = \frac{1}{\lambda\mu_0}$$

$$\therefore \lambda = \sqrt{\frac{\epsilon_0}{\mu_0}}$$

The current density in the sheet necessary to support such a 'surface wave' is obtained by applying the boundary condition

$$\mathbf{J} = \mathbf{n} \times \mathbf{H} = \mathbf{k} \times \mathbf{H} = \mathbf{i}\, \lambda A \sin n(x - ct)$$

1.6. In the case of no time variation the equation for **A** reduces to

$$\nabla^2 \mathbf{A} = -\mu \mathbf{J}$$

If we now split **A** and **J** into their cartesian components, the vector equation above becomes three equations of the form

$$\nabla^2 A_x = -\mu J_x \text{ .... etc.}$$

Now, Poisson's equation for the electrostatic potential ϕ in terms of the charge density ρ is just

$$\nabla^2 \phi = -\frac{1}{\epsilon}\rho$$

which is formally (that is to say, mathematically) the same equation as those for A_x, A_y and A_z.

The solution of Poisson's equation can be obtained by superposition of the potentials of elementary volumes as

$$\phi(x, y, z) = \frac{1}{4\pi\epsilon} \int_{v'} \frac{\rho(x', y', z')}{r} \, dv'$$

when r is the distance between the points (x, y, z) and (x', y', z').

The solution for (say) A_x can be written down immediately just by changing the notation. Thus, ϕ becomes A_x, ρ becomes J_x and $1/\epsilon$ becomes μ. Hence the solution for A_x is

$$A_x(x, y, z) = \frac{\mu}{4\pi} \int_{v'} \frac{J_x(x', y', z')}{r} \, dv'$$

with parallel solutions for A_y and A_z. If these solutions are now added to form a vector A then we have the final form

$$\mathbf{A}(x, y, z) = \frac{\mu}{4\pi} \int_{v'} \frac{\mathbf{J}(x', y', z')}{r} \, dv'$$

(Note that the dashed notation refers to source points and the undashed to field points.)

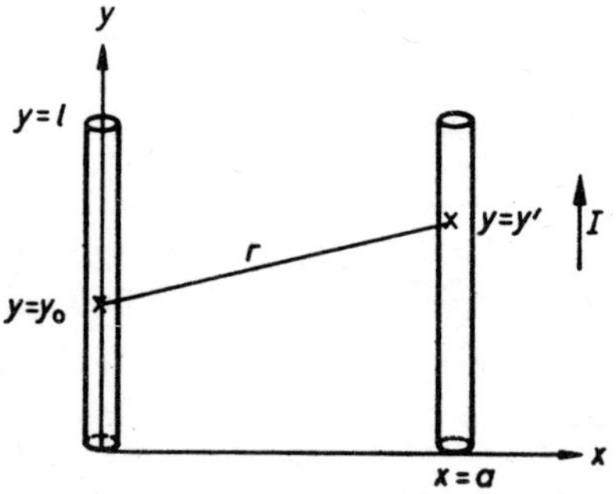

Figure 1.4.

For the second part consider the geometry of *Figure 1.4.* If Δ is the cross-sectional area (assumed small) of the wire then the current density in the second wire is

$$\mathbf{J}(y') = \mathbf{j}\frac{I}{\Delta}$$

and the vector potential is then given by

$$\mathbf{A} = \mathbf{j}\frac{\mu}{4\pi} \cdot \frac{I}{\Delta} \int_{v'} \frac{dv'}{r}$$

Since $r = \sqrt{\{a^2 + (y - y_0)^2\}}$ is independent of x' and z' the integration with respect to these two variables gives merely the cross-sectional area Δ and hence

$$\mathbf{A} = \mathbf{j}\, A(y_0) = \mathbf{j}\frac{\mu I}{4\pi} \int_0^l \frac{dy'}{\sqrt{\{(y' - y_0)^2 + a^2\}}}$$

which, on integration, gives the required answer,

$$A(y_0) = \frac{\mu I}{4\pi} \left\{ \sinh^{-1}\left(\frac{y_0}{a}\right) - \sinh^{-1}\left(\frac{y_0 - l}{a}\right) \right\}$$

In order to calculate the mutual inductance we note that, since $\mathbf{B} = \nabla \times \mathbf{A}$,

$$\int_S \mathbf{B} \, . \, d\mathbf{S} = \int_S \nabla \times \mathbf{A} \, . \, d\mathbf{S} = \oint_C \mathbf{A} \, . \, d\mathbf{l}$$

Consequently the flux linked with a circuit can be calculated from the line integral of the vector potential around the circuit. In our particular case we have only a limited circuit and we can consider the line integral to extend only over wire l. The total flux Ψ is

$$\Psi = \int_0^l A(y_0) \, dy_0$$

whence the mutual inductance M is

$$M = \frac{\Psi}{I} = \frac{1}{I} \int_0^l A(y_0) \, dy_0$$

The student may care to evaluate this.

[Note: $\int \sinh^{-1} \theta d\theta = \theta \sinh^{-1} \theta - \sqrt{(1 + \theta^2)}$.]

1.7. Maxwell's equations are

$$\nabla \times \mathbf{E} = -\frac{\partial \mathbf{B}}{\partial t}$$

$$\nabla \times \mathbf{H} = \mathbf{J} + \frac{\partial \mathbf{D}}{\partial t}$$

Noting that $\oint \mathbf{E} \times \mathbf{H} \,.\, d\mathbf{S} = \int \nabla . (\mathbf{E} \times \mathbf{H})\, dv$ we are led to consider the vector relation

$$\nabla . (\mathbf{E} \times \mathbf{H}) = \mathbf{H} . \nabla \times \mathbf{E} - \mathbf{E} . \nabla \times \mathbf{H}$$

$$= -\mathbf{H} . \frac{\partial \mathbf{B}}{\partial t} - \mathbf{E} . \left[\mathbf{J} + \frac{\partial \mathbf{D}}{\partial t}\right]$$

$$= -\mathbf{E} . \mathbf{J} - \frac{\partial}{\partial t} [\tfrac{1}{2}\mathbf{H} . \mathbf{B} + \tfrac{1}{2}\mathbf{E} . \mathbf{D}]$$

whence,

$$\oint \mathbf{E} \times \mathbf{H} . d\mathbf{S} = -\int \mathbf{E} . \mathbf{J}\, dv - \frac{\partial}{\partial t}\left[\int \tfrac{1}{2}\mathbf{H} . \mathbf{B}\, dv + \int \tfrac{1}{2}\mathbf{E} . \mathbf{D}\, dv\right]$$

Looking at the bracketed terms first of all, the quantities ½**H . B** and ½**E . D** are interpreted as the density of energy stored in the magnetic and electric fields respectively. The interpretation of the quantity **E . J** depends to some degree on the nature of **J**. If this is just conduction current (σ**E**) then **E . J** is the power loss per unit volume due to dissipation. If, on the other hand, **J** is present due to an impressed source, then −**E . J** represents the supplied power per unit volume, i.e. it is a measure of the work done on the system. Thus the surface integral is equal to the rate of *decrease* of stored energy, *minus* the power lost in dissipation *plus* the power supplied (or the rate of work done) from outside the system. The integral can therefore be interpreted as the flow of power *out* of the volume (or through the surface).

We can now consider the second part. Let z be the distance apart of the plates at time t. We can write

$$z = h(1 + t); \quad \dot{z} = h$$

If fringing is neglected the electric field E is

$$E = V/z \quad \text{whence} \quad D = \epsilon V/z$$

The displacement current density $\mathrm{d}D/\mathrm{d}t$ is therefore

$$\frac{\mathrm{d}D}{\mathrm{d}t} = -\frac{\epsilon V}{z^2}\,\dot{z} = -\frac{\epsilon Vh}{z^2}$$

If we now apply Ampère's law to a circle of radius r we find the circumferential magnetic field H is given by

$$H = -\frac{\epsilon Vh}{2z^2}\,r$$

At the outside edge of the capacitor $r = d/2$. Thus

$$\int \mathbf{E} \times \mathbf{H} \,.\, \mathrm{d}\mathbf{S} = \frac{V}{z} \cdot \frac{\epsilon Vhd}{2z^2} \cdot \pi zd = \frac{\pi\epsilon V^2hd^2}{4z^2}$$

Now the stored energies are

$$\int \tfrac{1}{2}\mathbf{E} \,.\, \mathbf{D}\,\mathrm{d}v = \frac{\pi\epsilon V^2d^2}{8z} \quad \text{and} \quad \int \tfrac{1}{2}\mathbf{H} \,.\, \mathbf{B}\,\mathrm{d}v = \frac{\pi\mu\epsilon^2V^2h^2d^2}{256z^3}$$

and the magnetic energy is negligible [assuming $h \ll c = (1/\sqrt{\mu\epsilon})$]. The rate of decrease of stored energy is then

$$-\frac{\mathrm{d}}{\mathrm{d}t}\left\{\int \tfrac{1}{2}\mathbf{E} \,.\, \mathbf{D}\,\mathrm{d}v\right\} = \frac{\pi\epsilon V^2hd^2}{8z^2}$$

There is clearly no dissipation as such but the plates of the capacitor are pulled apart and work is therefore being done on the system. For a distance z the force between the plates is

$$\frac{\pi d^2}{4} \cdot \frac{D^2}{2\epsilon} \cdot = \frac{\pi\epsilon V^2d^2}{8z^2}$$

and in a time $\mathrm{d}t$ the plates move by a distance $h\,\mathrm{d}t$. Thus

$$\text{work done/second} = \frac{\pi\epsilon V^2hd^2}{8z^2}$$

This quantity added to the rate of decrease of stored energy is equal to the surface integral evaluated at the edge of the capacitor, as the Poynting theorem requires.

The student should observe that the surface integral applies to a *closed* surface. We should strictly evaluate the contributions to the integral from the surfaces of the plates (which close the cylindrical surface). Over these, however, **E** x **H** . d**S** is zero and therefore this part of the integral is zero.

1.8. Since $\nabla \times \mathbf{E} = 0$ the field **E** can be written as the gradient of a scalar field (the negative gradient to maintain convention). That is,

$$\mathbf{E} = -\nabla\phi$$

Substituting this in the second equation gives us Poisson's equation

$$\nabla^2\phi = -\frac{\rho}{\epsilon}$$

The magnetic field equations under static conditions are

$$\nabla \times \mathbf{H} = \mathbf{J} \qquad \nabla \cdot \mathbf{B} = 0$$

Because the divergence of **B** is zero then the vector field **B** may be written as the curl of another vector field **A**. Hence

$$\mathbf{B} = \nabla \times \mathbf{A}$$

and, substituting,

$$\nabla \times (\nabla \times \mathbf{A}) = \mu\mathbf{J}$$

Since, from vector analysis,

$$\nabla \times (\nabla \times \mathbf{A}) = \nabla(\nabla \cdot \mathbf{A}) - \nabla^2\mathbf{A}$$

we obtain

$$\nabla^2\mathbf{A} = -\mu\mathbf{J} + \nabla(\nabla \cdot \mathbf{A})$$

Now only the curl of the field **A** is specified and by Helmholtz's theorem a vector field is specified uniquely only when both its divergence and curl are specified. There is thus a degree of freedom in a choice of **A**, its divergence we can choose as we please. We therefore choose

$$\nabla \cdot \mathbf{A} = 0$$

whence

$$\nabla^2 \mathbf{A} = -\mu \mathbf{J}$$

The rest of this problem is tackled in the solution to Problem **8.5**.

1.9. We can show that the given function satisfies Laplace's equation by substituting it into the equation. Since ϕ is zero when $z = L$ and when $r = b$ it is suitable for the region with conductors (i.e. equipotential surfaces) stated.

Inside the resistor of *Figure 1.1* only E_z exists and is equal to V/L in magnitude. If we assume that the potential function given represents the potential in the air region then, in this region, E_z is given by

$$E_z = -\frac{\partial \phi}{\partial z} = \frac{A}{L} \ln\left(\frac{r}{b}\right)$$

Since E_z must be continuous across the boundary $r = a$ we have

$$\frac{A}{L} \ln \frac{a}{b} = \frac{V}{L}$$

whence
$$A = \frac{V}{\ln (a/b)}$$

Thus, in the air region,

$$E_r = -\frac{\partial \phi}{\partial r} = -\frac{V\left(1 - \dfrac{z}{L}\right)}{r \ln (a/b)}$$

$$E_z = -\frac{\partial \phi}{\partial z} = \frac{V \ln (r/b)}{L \ln (a/b)}$$

The magnetic field has only one component H_θ which can easily be calculated by applying Ampère's law to a circle of radius r. Thus

$$2\pi r H_\theta = \pi a^2 \sigma E_z = \frac{\pi a^2 \sigma V}{L}$$

whence
$$H_\theta = \frac{\sigma a^2 V}{2L} \frac{1}{r}$$

The power flowing into the resistive cylinder can be calculated from the component of the Poynting vector in the direction $-\mathbf{a}_r$,(at $r = a$). This gives the total power flow *into* the cylinder as

$$\int [-\mathbf{a}_r \cdot \mathbf{E} \times \mathbf{H}]_{r=a} \, ds = 2\pi a \int_{z=0}^{L} [E_z H_\theta]_{r=a} \, dz = \frac{\pi a^2 V^2 \sigma}{L}$$

Now the resistance of a resistor of radius a, length L and conductivity σ is just $R = L/(\pi a^2 \sigma)$. The power loss is $V^2/R = (\pi a^2 V^2 \sigma)/L$ as computed from the Poynting vector.

1.10. The importance of this problem rests solely on the fact that many students seem to have difficulty in understanding the nature of phasor vectors as used in field theory.

If we look at the quoted expression for **E** we see that it consists of two parts, one proportional to $\exp(j\omega t)$ and the other proportional to $\exp(-j\omega t)$. The other field vectors will have similar parts as will the scalar quantities. Applying the principle of superposition, each of these parts can be considered separately. Let us look at the first, therefore. Substituting this form into Faraday's law, for example, we obtain

$$\nabla \times (\tfrac{1}{2}\hat{\mathbf{E}} \exp j\omega t) = -\frac{\partial}{\partial t}(\tfrac{1}{2}\hat{\mathbf{B}} \exp j\omega t) = -\tfrac{1}{2} j\omega \hat{\mathbf{B}} \exp j\omega t$$

and since the curl operator is independent of time

$$\nabla \times \hat{\mathbf{E}} = -j\omega \hat{\mathbf{B}}$$

Doing the same for all of the original Maxwell's equations gives us the set of phasor Maxwell's equations

$$\nabla \times \hat{\mathbf{E}} = -j\omega \hat{\mathbf{B}}\,; \qquad \nabla \cdot \hat{\mathbf{D}} = \hat{\rho}$$

$$\nabla \times \hat{\mathbf{H}} = \hat{\mathbf{J}} + j\omega \hat{\mathbf{D}}\,; \qquad \nabla \cdot \hat{\mathbf{B}} = 0$$

Now the other part of the field vector is just the complex conjugate of the part we have already considered. The solutions are thus the complex conjugate set to the solutions of the phasor equations above. In practice, therefore, we need only to consider the phasor equations themselves.

The three items of the last part of the problem are all solved by the same method and we will only consider the time average stored electric energy density. In this case we *cannot* apply superposition since energy (and power) are not linear functions.

The stored energy in the electric field is

$$U_e = \tfrac{1}{2}\mathbf{E} \cdot \mathbf{D} = \tfrac{1}{8}[\hat{\mathbf{E}}\, e^{j\omega t} + \hat{\mathbf{E}}^* e^{-j\omega t}] \cdot [\hat{\mathbf{D}}\, e^{j\omega t} + \hat{\mathbf{D}}^* e^{-j\omega t}]$$
$$= \tfrac{1}{8}[\hat{\mathbf{E}} \cdot \hat{\mathbf{D}}^* + \hat{\mathbf{E}}^* \cdot \hat{\mathbf{D}} + \hat{\mathbf{E}} \cdot \hat{\mathbf{D}} \exp(2j\omega t) + \hat{\mathbf{E}}^* \cdot \hat{\mathbf{D}}^* \exp(-2j\omega t)]$$

The first two terms are independent of time and the second two may be combined in the form

$$2\,|\,\hat{\mathbf{E}}\,|\,|\,\hat{\mathbf{D}}\,|\cos(2\omega t + \theta)$$

where θ is some phase angle.

Now the time average value of $\cos(2\omega t + \theta)$ over one cycle is zero and since $\hat{\mathbf{D}} = \epsilon\hat{\mathbf{E}}$, the time average stored energy is given by

$$\langle U_e \rangle = \tfrac{1}{8}[\hat{\mathbf{E}} \cdot \hat{\mathbf{D}}^* + \hat{\mathbf{E}}^* \cdot \hat{\mathbf{D}}] = \tfrac{1}{4}\epsilon\,|\,\hat{E}\,|^2$$

In an identical fashion we can show that the time average stored magnetic energy is

$$\langle U_m \rangle = \tfrac{1}{4}\mu\,|\,\hat{H}\,|^2$$

and that the time average power flow vector is

$$\langle \mathbf{S} \rangle = \tfrac{1}{2}\mathrm{Re}\{\hat{\mathbf{E}} \times \hat{\mathbf{H}}^*\}$$

The complex vector $\tfrac{1}{2}\hat{\mathbf{E}} \times \hat{\mathbf{H}}^*$ is known as the complex Poynting vector and the real part of this gives the power flow.

CHAPTER 2

2.1 (*a*) Using the method of solution known as the 'separation of variables', we assume a solution exists which can be written as a product of functions of a single coordinate, i.e. in the rectangular coordinate system of our problem we assume a solution of the form:

$$\phi = X(x)\,Y(y)$$

Substituting this into Laplace's equation gives

$$Y\frac{d^2X}{dx^2} + X\frac{d^2Y}{dy^2} = 0$$

which may be written in the form

$$\frac{1}{X}\frac{d^2X}{dx^2} = -\frac{1}{Y}\frac{d^2Y}{dy^2}$$

Clearly the L.H.S. of this equation is unaffected by a change in the value of y and similarly the R.H.S. is not affected by a change in x. Both sides then must be equal to a constant, for instance k^2, the separation constant. Hence

$$\frac{1}{X}\frac{d^2X}{dx^2} = -k^2, \qquad \frac{1}{Y}\frac{d^2Y}{dy^2} = k^2.$$

k can be real or complex in general, but we must decide which for our problem. Choosing k real gives rise to sinusoids in the x direction and these can satisfy the boundary conditions imposed in our problem, i.e. that at $x = 0, a$; $\phi = 0$. If we had chosen k imaginary, then we would not have obtained periodic solutions in x and therefore could not have satisfied the boundary conditions.

$$\therefore \qquad X = A_k \cos kx + B_k \sin kx$$

Since $\phi = 0$ at $x = 0$ for all y, $A_k = 0$.
Also, $\phi = 0$ at $x = a$ for all y, and this limits k to the values $p\pi/a$ where p is an integer.

$$\therefore \qquad X = B_k \sin\left(p\frac{\pi}{a}x\right).$$

The corresponding solution for Y can either be written in terms of exponential or hyperbolic functions but since we have to satisfy the boundary condition that $\phi \to 0$ as $y \to \infty$ then it is more convenient for us to choose the former.

i.e.

$$Y = C_k e^{ky} + D_k e^{-ky}$$

C_k must be zero since this function grows with increasing y and therefore cannot satisfy the boundary conditions. Hence we see that a form of solution that satisfies Laplace's equation and the boundary conditions at $x = 0, a; y = \infty$, is

$$\phi = F_p \sin\left(p\frac{\pi}{a}x\right) e^{-p(\pi/a)y}$$

where F_p is $B_k D_k$.

Finally we must satisfy the remaining condition at $y = 0$. By inspection we see that the above solution by itself cannot do this. Because the equations are linear, however, we can superpose solutions of the form above over all possible values of the parameter p if we wish, i.e.

$$\phi = \sum_{p=1}^{\infty} F_p \sin\left(p\frac{\pi}{a}x\right) e^{-p(\pi/a)y}$$

We may now regard F_p as the coefficients of the Fourier sine expansion of the given potential along the $y = 0$ axis. To determine the values of these coefficients we make use of the orthogonality of the sine functions,

$$\int_0^{a/2} x \sin\left(p\frac{\pi}{a}x\right) dx + \int_{a/2}^{a} (a - x) \sin\left(p\frac{\pi}{a}x\right) dx = F_p \int_0^a \sin^2\left(p\frac{\pi}{a}x\right) dx$$

$$= F_p \frac{a}{2}$$

All of the other terms vanish because

$$\int_0^a \sin\left(m\frac{\pi}{a}x\right) \sin\left(n\frac{\pi}{a}x\right) dx = 0 \qquad (m \neq n)$$

Carrying out the indicated integrations we obtain

$$\int_0^{a/2} x \sin\left(p\frac{\pi}{a}x\right) dx = \left(\frac{a}{p\pi}\right)^2 \sin\left(p\frac{\pi}{2}\right) - \frac{a^2}{2p\pi}\cos\left(p\frac{\pi}{2}\right)$$

$$\int_{a/2}^{a} x \sin\left(p\frac{\pi}{a}x\right) dx = -\left(\frac{a}{p\pi}\right)^2 \sin\left(p\frac{\pi}{2}\right) - \frac{a^2}{p\pi}\cos(p\pi) + \frac{a^2}{2p\pi}\cos\left(p\frac{\pi}{2}\right)$$

$$\int_{a/2}^{a} a \sin\left(p\frac{\pi}{a}x\right) dx = -\left(\frac{a^2}{p\pi}\right)\left[\cos(p\pi) - \cos\left(p\frac{\pi}{2}\right)\right]$$

Hence
$$F_p = \frac{4a}{\pi^2 p^2}\sin\left(p\frac{\pi}{2}\right)$$

The sine series then will contain only odd values of p and every other term will be of alternating sign beginning with *a* +.

$$\therefore \qquad \phi = \frac{4a}{\pi^2}\sum_{n=0}^{\infty}\frac{(-1)^n}{(2n+1)^2}\sin\left\{(2n+1)\frac{\pi x}{a}\right\} e^{-(2n+1)(\pi y/a)}$$

$p = (2n + 1) - n$ integer.

(*b*) We see from the boundary conditions that the current density along the y axis is perpendicular to the boundary and its value will be

$$-\frac{1}{\rho}\frac{\partial\phi}{\partial x}\bigg|_{x=0}$$

where ρ is the surface resistivity of the material.

Integrating this current density along the entire y axis will, by symmetry, give us half of the total current in the strip, i.e.

$$I = 2\int_0^\infty \left(-\frac{1}{\rho} \frac{\partial \phi}{\partial x}\bigg|_{x=0} \right) \mathrm{d}y$$

$$= -\frac{8a}{\rho\pi^2} \int_0^\infty \sum_{n=0}^\infty \frac{(-1)^n}{(2n+1)^2} (2n+1)\frac{\pi}{a}\, \mathrm{e}^{-(2n+1)(\pi y/a)}\, \mathrm{d}y$$

$$= \frac{8a}{\rho\pi^2} \sum_{n=0}^\infty \frac{(-1)^n}{(2n+1)^2}\, \mathrm{e}^{-(2n+1)(\pi/a)y} \bigg|_{y=0}^{y=\infty}$$

i.e.
$$I = -\frac{8a}{\rho\pi^2} \sum_{n=0}^\infty \frac{(-1)^n}{(2n+1)^2}$$

Let us consider the infinite series; the first few terms are:

$$1 - \frac{1}{9} + \frac{1}{25} - \frac{1}{49} + \frac{1}{81} - \ldots$$

$$\doteqdot \underbrace{1 - 0{\cdot}111 + 0{\cdot}04}\quad 0{\cdot}02 + 0{\cdot}01 \quad \ldots$$

The terms decrease monotonically and alternate in sign hence the series is convergent. We can estimate the sum of the series from the terms above, i.e.

$$0{\cdot}929 \quad 0{\cdot}02 + 0{\cdot}01 > \text{sum} > 0{\cdot}929 - 0{\cdot}02$$

$\therefore$
$$0{\cdot}919 > \text{sum} > 0{\cdot}909$$

hence,
$$\text{sum} = 0{\cdot}914 \pm 0{\cdot}005$$

This is an accuracy of the order of ½% and is adequate for our purpose.

$\therefore$
$$I = \frac{8}{1000} \cdot \frac{0{\cdot}1}{\pi^2} \cdot 0{\cdot}914\ \text{A}$$

$\therefore$
$$I = 74\ \mu\text{A}$$

2.2. Following the same reasoning as in the previous problem we choose a solution having sinusoidal variations in the y direction and hyperbolic variations in the x direction, i.e.

$$V = (A_k \cos ky + B_k \sin ky)(C_k \cosh kx + D_k \sinh kx)$$

when $x = 0$, $V = 0$ for all $y \therefore C_k = 0$

when $y = 0$, $V = 0$ for all $x \therefore A_k = 0$

$\therefore \qquad V = F_k \sinh kx \sin ky \qquad\qquad F_k \equiv B_k D_k$

when $y = b$, $\dfrac{\partial V}{\partial y} = 0$ for all x

$\therefore \qquad k = (2n+1)\dfrac{\pi}{2b}$, n integer, 0, 1, 2 . . . . etc.

Again a single term will not satisfy the remaining boundary condition so we employ an infinite series of them.

$$V = \sum_{n=0}^{\infty} F_n \sinh\{(2n+1)\pi x/2b\} \sin\{(2n+1)\pi y/2b\}$$

when $x = a$, $V = V_0 = \displaystyle\sum_{n=0}^{\infty} F_n \sinh\{(2n+1)\pi a/2b\} \sin\{(2n+1)\pi y/2b\}$

If we now expand V_0 as a half range Fourier sine series where the half range is the interval $0 \rightarrow 2b$, we can equate the coefficients to

$$F_n \sinh\{(2n+1)\pi a/2b\}$$

The fact that the potential has not been defined in the range $b \rightarrow 2b$ need not concern us as we can choose any function we like for it and it will not affect the solution in the region of interest, i.e. within the rectangle; the uniqueness theorem assures us of that.

$$V_0 = \sum_{s=1}^{\infty} M_s \sin\left(\frac{s\pi y}{2b}\right)$$

$$M_s = \frac{2}{2b}\int_0^{2b} V_0 \sin\left(\frac{s\pi y}{2b}\right) dy = \frac{4V_0}{s\pi} \qquad (s \text{ odd})$$

$$= 0 \qquad (s \text{ even})$$

$$\therefore \qquad V = \sum_{n=0}^{\infty}\left(\frac{4V_0}{\pi}\right)\frac{\sinh\{(2n+1)\pi x/2b\}\sin\{(2n+1)\pi y/2b\}}{(2n+1)\sinh\{(2n+1)\pi a/2b\}}$$

2.3. Consider the equation

$$\frac{\partial^2 V}{\partial r^2} + \frac{1}{r}\frac{\partial V}{\partial r} + \frac{\partial^2 V}{\partial z^2} = 0$$

Using the technique of separation of variables we assume a solution exists of the form

$$V = R(r)\, Z(z)$$

Substituting this into the above equation and rearranging:

$$\frac{1}{R}\frac{d^2R}{dr^2} + \frac{1}{Rr}\frac{dR}{dr} = -\frac{1}{Z}\frac{d^2Z}{dz^2}$$

Again, because the L.H.S. does not contain z it cannot be affected by a change in this independent variable. A similar argument is applicable to the R.H.S. as regards the variable r. For these two functions to be equal then they must equal the same constant, p^2 say.

p can be real or complex in general but because of the form of solution desired we shall choose the following with p^2 positive.

$$\frac{d^2Z}{dz^2} - p^2 Z = 0, \quad r^2\frac{d^2R}{dr^2} + r\frac{dR}{dr} + p^2r^2 R = 0$$

The first of these gives a solution in hyperbolics whilst the second, being a Bessel differential equation of zero order, yields solutions involving zero order Bessel functions of the first and second kind. Our complete solution is

$$V = \{A\,J_0(pr) + BY_0(pr)\}\{C\cosh pz + D\sinh pz\}$$

We now apply the boundary conditions imposed by our particular problem. V being finite as $r \to 0$ implies $B = 0$. This must be so since Y_0 is that type of solution which increases without limit as r tends to zero. $V = 0$ when $r = a$, restricts the separation constant to discrete values determined by the equation $J_0\,(pa) = 0$, i.e. $p = (\lambda_i/a)$ where λ_i are the positive roots of this equation. $V = 0$ when $z = 0$ implies $C = 0$.

With the above conditions imposed our solution to date has taken the form

$$V = F\sinh\left(\frac{\lambda_i}{a}z\right)J_0\left(\frac{\lambda_i}{a}r\right),\ F \equiv A\,D$$

This by itself will not satisfy the remaining boundary condition at $z = h$. We now consider using an infinite number of such solutions, i.e. let

$$V = \sum_{i=1}^{\infty} F_i \sinh\left(\frac{\lambda_i}{a}z\right)J_0\left(\frac{\lambda_i}{a}r\right)$$

When $z = h$, $$V = \sum_{i=1}^{\infty} F_i \sinh\left(\frac{\lambda_i}{a}h\right)J_0\left(\frac{\lambda_i}{a}r\right) = E(a^2 - r^2)$$

We can now equate F_i sinh $[(\lambda_i/a)h]$ to the coefficients, G_i say, of the Fourier-Bessel expansion of $E\,(a^2 - r^2)$.

To determine these coefficients, G_i, we make use of the orthogonality properties of the Bessel functions in much the same way as we do with the sinusoids in a Fourier series. The required integrals are

$$\int_0^a G_i\, r\, J_0{}^2\left(\frac{\lambda_i}{a}r\right)dr = G_i\frac{a^2}{2}J_1{}^2(\lambda_i)$$

$$\int_0^a G_i\, r\, J_0\left(\frac{\lambda_m}{a}r\right)J_0\left(\frac{\lambda_i}{a}r\right)dr = 0 \qquad (m \neq i)$$

Hence
$$G_i = \frac{2}{a^2 J_1^2(\lambda_i)} \left\{ \int_0^a E\,(a^2 - r^2)\, r\, J_0\!\left(\frac{\lambda_i}{a} r\right) dr \right\}$$

Performing the indicated integrals we obtain:

$$\int_0^a (E\,a^2)\, r\, J_0\!\left(\frac{\lambda_i}{a} r\right) dr = \frac{Ea^4}{\lambda_i^2} \int_0^{\lambda_i} v\, J_0(v)\, dv$$

$$= \frac{Ea^4}{\lambda_i} J_1(\lambda_i)$$

$$\int_0^a E\, r^3\, J_0\!\left(\frac{\lambda_i}{a} r\right) dr = E \left\{ r^2 \int r\, J_0\!\left(\frac{\lambda_i}{a} r\right) dr \Bigg|_0^a \right.$$

$$\left. - \int_0^a 2r \left(\int r\, J_0\!\left(\frac{\lambda_i}{a} r\right) dr \right) dr \right\}$$

The first of these is similar to the one above and gives

$$\frac{a^4}{\lambda_i} J_1(\lambda_i)$$

Finally
$$-\int_0^a 2r \left(\int r\, J_0\!\left(\frac{\lambda_i}{a} r\right) dr \right) dr$$

$$= -2 \left(\frac{a}{\lambda_i}\right)^4 \int_0^{\lambda_i} v \left(\int v\, J_0(v)\, dv \right) dv$$

$$= -2 \frac{a^4}{\lambda_i^4} \int_0^{\lambda_i} v^2\, J_1(v)\, dv$$

$$= -2 \frac{a^4}{\lambda_i^4} v^2 J_2(v) \Bigg|_0^{\lambda_i} = -2 \frac{a^4}{\lambda_i^2} J_2(\lambda_i)$$

$$\therefore \qquad G_i = \frac{2}{a^2 J_1^2(\lambda_i)} \left\{ 2 \frac{a^4}{\lambda_i^2} E\, J_2(\lambda_i) \right\}$$

But
$$\frac{2J_1(\lambda_i)}{\lambda_i} = J_2(\lambda_i) + J_0(\lambda_i) = J_2(\lambda_i)$$

$$\therefore \qquad G_i = \frac{8a^2 E}{\lambda_i^3 J_1(\lambda_i)} = F_i \sinh\left(\frac{\lambda_i}{a} h\right)$$

$$\therefore \qquad V = 8a^2 E \sum_{i=1}^{\infty} \frac{1}{\lambda_i^3 J_1(\lambda_i)} \cdot \frac{\sinh\left(\frac{\lambda_i}{a} z\right)}{\sinh\left(\frac{\lambda_i}{a} h\right)} J_0\left(\frac{\lambda_i}{a} r\right)$$

2.4. Let us assume that a solution can be found by separating the variables, i.e. assume

$$\phi = R(r)\,\Theta(\theta)$$

Substituting this into the equation and rearranging the terms gives

$$\frac{r^2}{R}\frac{d^2R}{dr^2} + \frac{r}{R}\frac{dR}{dr} = -\frac{1}{\Theta}\frac{d^2\Theta}{d\theta^2} = n^2$$

where n is the separation constant.

From a consideration of the boundary conditions we can select the appropriate sign for n^2; in this case we choose n^2 to be positive. Hence

$$\Theta = A_n \sin(n\theta) + B_n \cos(n\theta)$$

The symmetry of the problem tells us that the cosine terms are not allowable so we set $B_n = 0$.

Now the equation governing the radial dependence is

$$r^2 \frac{d^2R}{dr^2} + r\frac{dR}{dr} = n^2 R$$

and we might try a solution of the form $R = r^p$ and we find by substituting this back into the differential equation that $p = \pm n$.

In this problem, the negative exponent term is not admissible since this would make the potential infinite at the origin.

Hence
$$\phi = \sum_{n=1}^{\infty} A_n r^n \sin(n\theta)$$

When $r = a$,
$$\phi = \sum_{n=1}^{\infty} A_n a^n \sin(n\theta)$$

hence we can interpret $(A_n a^n)$ as the coefficients of the half range Fourier sine series for the potential on the cylinder, i.e.

$$A_n a^n = \frac{2}{\pi}\int_0^{\pi} E \sin(n\theta)\, d\theta = \frac{4E}{n\pi} \qquad (n \text{ odd})$$

$$= 0 \qquad (n \text{ even})$$

$$\therefore \qquad \phi = \frac{4E}{\pi} \sum_{\text{all } n \text{ odd}} \left(\frac{r}{a}\right)^n \frac{\sin(n\theta)}{n}$$

or, this may be put in the more convenient form

$$\phi = \frac{4E}{\pi} \sum_{n=0}^{\infty} \left(\frac{r}{a}\right)^{2n+1} \frac{\sin(2n+1)\theta}{(2n+1)}$$

2.5. Since there are no currents involved we can employ the scalar magnetic potential, ϕ, thus reducing the problem essentially to the solution of the two dimensional Laplace's equation in plane polar coordinates subject to the following boundary conditions:

(*i*) $\phi_3 \to -H_0\, r \cos\theta$ (see *Figure 2.3*)

(*ii*) when $r = b$, $\phi_2 = \phi_3$,

(*iii*) when $r = a$, $\phi_1 = \phi_2$,

(*iv*) when $r = b$, $\mu \dfrac{\partial \phi_2}{\partial r} = \dfrac{\partial \phi_3}{\partial r}$

(*v*) when $r = a$, $\dfrac{\partial \phi_1}{\partial r} = \mu \dfrac{\partial \phi_2}{\partial r}$

(*vi*) ϕ_1 finite as $r \to 0$.

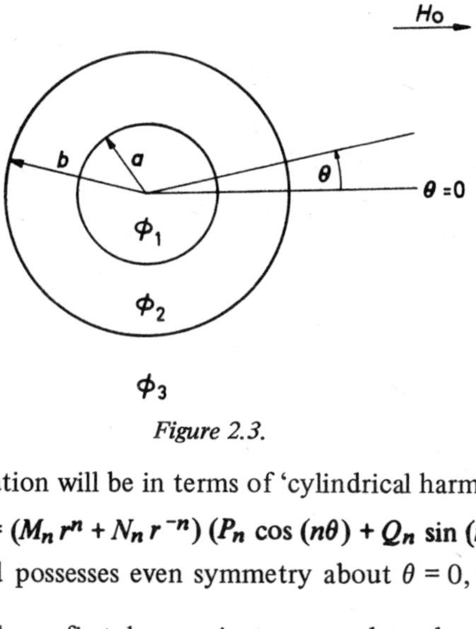

Figure 2.3.

Clearly our solution will be in terms of 'cylindrical harmonics', i.e.

$$\phi = (M_n r^n + N_n r^{-n})\,(P_n \cos(n\theta) + Q_n \sin(n\theta))$$

and since the field possesses even symmetry about $\theta = 0$, there can be no sine terms present.

Now ϕ_3 must have first harmonic terms and tend to $-H_0 r\cos\theta$ at large values of r, therefore we will choose a potential of the form:

$$\phi_3 = \left(\frac{A}{r} - H_0 r\right)\cos\theta$$

hence

$$\phi_2 = \left(\frac{B}{r} + Cr\right)\cos\theta$$

and

$$\phi_1 = Dr\cos\theta$$

where we have used conditions (*i*) and (*vi*).

We have four conditions yet to apply and four unknown constants, A, B, C and D, therefore we can solve our problem.

Using (*ii*)
$$\frac{B}{b} + Cb = \frac{A}{b} - H_0 b \qquad \ldots .(1)$$

Using (*iii*)
$$Da = \frac{B}{a} + Ca \qquad \ldots .(2)$$

Using (*iv*)
$$\mu\left[C - \frac{B}{b^2}\right] = \left[-\frac{A}{b^2} - H_0\right] \qquad \ldots .(3)$$

Using (*v*)
$$D = \mu\left[-\frac{B}{a^2} + C\right] \qquad \ldots .(4)$$

Before evaluating all of these constants let us look ahead and see what we wish to do with them.

We are asked to calculate the flux entering the cylinder and to compare this with what would enter the same region if the cylinder were removed. The flux entering the cylinder is simply

$$\Phi_m = 2\int_0^{\pi/2}\left[-\mu_0 \frac{\partial \phi_3}{\partial r}\bigg|_{r=b}\right] b\, d\theta$$

$$= 2b\,\mu_0\left[\frac{A}{b^2} + H_0\right]\int_0^{\pi/2} \cos\theta\, d\theta$$

$$\therefore \quad \frac{\Phi_m}{\Phi_0} = \frac{A}{b^2 H_0} + 1 \qquad \ldots .(5)$$

where $\Phi_0 = 2\mu_0 b H_0$ is the flux that would enter the region if the cylinder were removed. Our problem then is to determine A.

Eliminating D from Equations (2) and (4) we get

$$\frac{B}{C} = \frac{a^2(\mu - 1)}{(\mu + 1)}$$

Dividing Equation (1) by Equation (3) and rearranging terms,

$$\frac{A - H_0 b^2}{A + H_0 b^2} = \frac{B/C + b^2}{\mu[B/C - b^2]} = \frac{(A/H_0 b^2) - 1}{(A/H_0 b^2) + 1}$$

Substituting in for B/C and solving for $A/H_0 b^2$,

$$\frac{A}{H_0 b^2} = \frac{(b^2 - a^2)(\mu^2 - 1)}{b^2(\mu + 1)^2 - a^2(\mu - 1)^2}$$

Substituting into Equation (5) and simplifying we obtain the desired result:

$$\frac{\Phi_m}{\Phi_0} = \frac{2\mu[b^2(\mu + 1) - a^2(\mu - 1)]}{b^2(\mu + 1)^2 - a^2(\mu - 1)^2}$$

2.6. From the first part of this question we know that the stress on the surface of the sphere is $\sigma^2/2\epsilon_0$ per unit area, which written in terms of field quantities becomes $\epsilon_0 E_s{}^2/2$, where E_s is the electric field over the sphere. The problem then is to find E_s.

Because of the symmetry of the problem we shall solve Laplace's equation in spherical polar coordinates bearing in mind that the potential shall not vary with azimuthal angle, ψ, and that it is subjected to the following boundary conditions:

(*i*) as $r \to \infty$, $\phi \to -E_0 r \cos\theta$ (see *Figure 2.4(a)*)
(*ii*) at $r = a$, $\phi = 0$.

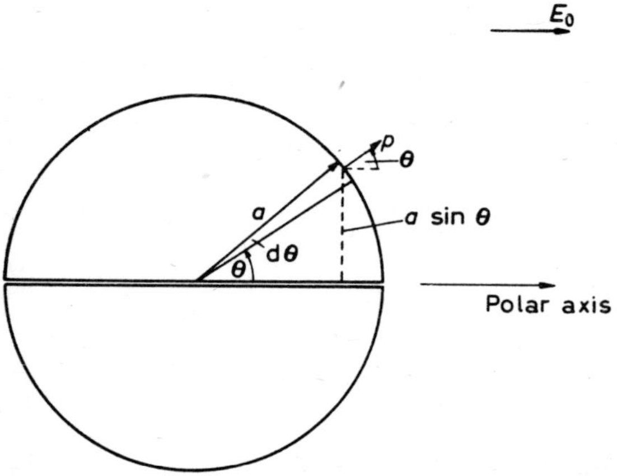

Figure 2.4(a).

Assuming a 'separation of variables' solution, $\phi = R(r)\,P(\theta)$ we substitute this into

$$\frac{1}{r^2}\frac{\partial}{\partial r}\left(r^2\frac{\partial\phi}{\partial r}\right) + \frac{1}{r^2 \sin\theta}\frac{\partial}{\partial\theta}\left(\sin\theta\frac{\partial\phi}{\partial\theta}\right) = 0$$

and after rearranging the terms obtain

$$\frac{1}{R}\frac{\mathrm{d}}{\mathrm{d}r}\left(r^2\frac{\mathrm{d}R}{\mathrm{d}r}\right) = -\frac{1}{P\sin\theta}\frac{\mathrm{d}}{\mathrm{d}\theta}\left(\sin\theta\frac{\mathrm{d}P}{\mathrm{d}\theta}\right) = \text{a constant}$$

By choosing the constant of the form $n(n + 1)$ where n is an integer, the separated equations become

$$\frac{1}{\sin\theta}\frac{\mathrm{d}}{\mathrm{d}\theta}\left(\sin\theta\frac{\mathrm{d}P}{\mathrm{d}\theta}\right) + n(n+1)P = 0$$

$$\frac{\mathrm{d}}{\mathrm{d}r}\left(r^2\frac{\mathrm{d}R}{\mathrm{d}r}\right) - n(n+1)R = 0$$

The first of these we recognise as Legendre's differential equation and it has solutions P_n $(\cos\theta)$ (the second solution is not admissible because it becomes infinite along the polar axis). It can readily be verified that the other equation has solutions of the form

$$R = A_n r^n + B_n r^{-(n+1)}$$

hence $$\phi_n = \{A_n r^n + B_n r^{-(n+1)}\}P_n$$

and we may superpose as many of these solutions as we require to fit the boundary conditions imposed in our problem.

Considering conditions (*i*), we must have terms in P_1, i.e.

$$\phi = \left(-E_0 r + \frac{B}{r^2}\right)\cos\theta$$

From (*ii*) $\phi = 0$ for all θ, when $r = a$

$$\therefore \qquad -E_0 a + \frac{B}{a^2} = 0$$

$$\therefore \qquad B = a^3 E_0$$

$$\phi = \left(-r + \frac{a^3}{r^2}\right) E_0 \cos\theta$$

$$E_s = -\left.\frac{\partial\phi}{\partial r}\right|_{r=a} = -\left(-1 - 2\frac{a^3}{r^3}\right) E_0 \cos\theta \Big|_{r=a}$$

$$= 3\,E_0 \cos\theta$$

Again, using the symmetry of the problem, we see that the resultant force on a hemisphere acts normal to its base and is directed so as to part the hemispheres.

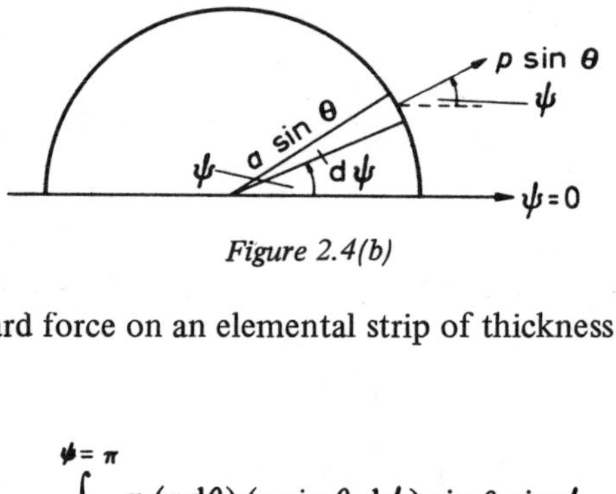

Figure 2.4(b)

Thus the upward force on an elemental strip of thickness $a\,d\theta$ [see *Figure 2.4(b)*] is

$$\int_{\psi=0}^{\psi=\pi} p\,(a\,d\theta)\,(a\sin\theta\,d\psi)\sin\theta\sin\psi$$

where $p = \dfrac{9}{2}\epsilon_0 E_0^{\,2}\cos^2\theta$

which equals $$pa^2\sin^2\theta\,d\theta\,[-\cos\psi]_0^\pi$$

$$= 2pa^2\sin^2\theta\,d\theta$$

$$\text{Hence total force on hemisphere} = \int_0^{\pi} 9\epsilon_0 E_0{}^2 a^2 \cos^2\theta \sin^2\theta \, d\theta$$

$$= \frac{9}{4}\epsilon_0 a^2 E_0{}^2 \int_0^{\pi} (\sin 2\theta)^2 \, d\theta$$

$$= \frac{9}{8}\epsilon_0 a^2 \pi E_0{}^2$$

$$= 4\pi\epsilon_0 \left(\frac{9}{32} a^2 E_0{}^2\right)$$

2.7. Using spherical polar coordinates, with the polar axis as shown in *Figure 2.5* parallel to the uniform field vector **E**, we will have solutions of the form

$$\phi_n = [A_n r^n + B_n r^{-(n+1)}]\, P_n$$

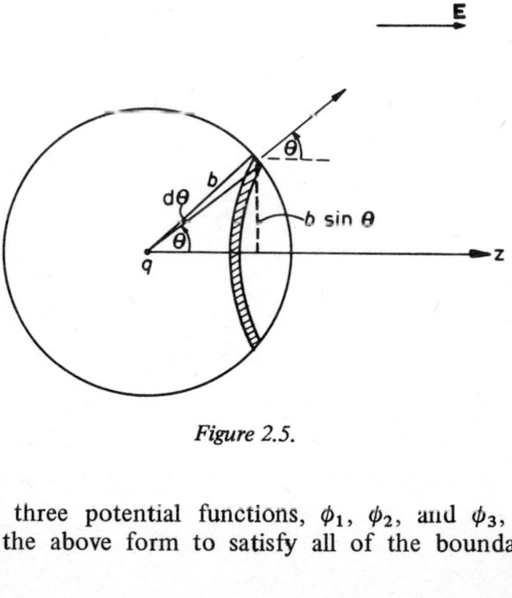

Figure 2.5.

and defining three potential functions, ϕ_1, ϕ_2, and ϕ_3, we superpose solutions of the above form to satisfy all of the boundary conditions.

These are:

(*i*) as $r \to \infty$, $\phi_3 \to -Er\cos\theta$

(*ii*) when $r = a$, $\phi_2 = \phi_3$

(*iii*) when $r = b$, $\phi_1 = \phi_2$

(*iv*) when $r = a$, $\dfrac{\partial\phi_3}{\partial r} = K\dfrac{\partial\phi_2}{\partial r}$,

(*v*) when $r = b$, $\dfrac{\partial\phi_1}{\partial r} = K\dfrac{\partial\phi_2}{\partial r}$,

(*vi*) ϕ_1 finite as $r \to 0$

From (*i*) we must have solutions with P_1 terms, i.e.

$$\phi_3 = \left(-Er + \frac{B}{r^2}\right)\cos\theta$$

$$\phi_2 = \left(Cr + \frac{D}{r^2}\right)\cos\theta$$

and using (*vi*)

$$\phi_1 = Fr\cos\theta$$

using (*ii*)

$$-Ea + \frac{B}{a^2} = Ca + \frac{D}{a^2} \qquad \ldots.(1)$$

using (*iii*)

$$Fb = Cb + \frac{D}{b^2} \qquad \ldots.(2)$$

using (*iv*)

$$K\left[C - \frac{2D}{a^3}\right] = -E - \frac{2B}{a^3} \qquad \ldots.(3)$$

using (v)

$$F = K\left[C - \frac{2D}{b^3}\right] \quad \ldots .(4)$$

We wish to determine F in terms of E. Therefore eliminating B from Equations (1) and (3) gives

$$-3a^3E = a^3(K + 2)\,C - 2D(K - 1) \quad \ldots .(5)$$

From Equations (2) and (4) we obtain

$$D = \frac{(K - 1)b^3F}{3K} \quad \ldots .(6)$$

$$C = \frac{(2K + 1)F}{3K} \quad \ldots .(7)$$

Substituting Equations (6) and (7) into (5) gives

$$F = -9a^3KE\,\{(K + 2)(2K + 1)\,a^3 - 2(K - 1)^2b^3\}^{-1}$$

From the form of the potential for the field in the empty interior region we see that the electric field is uniform and directed along the polar axis; the intensity of the field being F.

(a) If we now place a point charge q at the origin of coordinates, we can see by the symmetry that the electric field at the origin will be unchanged. Hence the force on the charge will be $-qF$.

(b) Without the charge q we see, again using symmetry, that there can be no net force on the dielectric shell.

Now for the purpose of calculating the force on the shell, we can replace the polarized dielectric by charge distributions $\nabla.\mathbf{P}$ throughout the volume of the dielectric and $\mathbf{P}\,.\,\mathbf{n}$ over the bounding surfaces. But

$$\mathbf{P} = (K - 1)\epsilon_0\mathbf{E}_2 = -(K - 1)\epsilon_0\nabla\phi_2$$

$$\therefore \qquad \nabla\cdot\mathbf{P} = -(K - 1)\epsilon_0\nabla^2\phi_2 = 0$$

hence there is no volume distribution within the homogeneous dielectric.

Let us consider the bound charge distribution over the surface of radius b. The outward normal vector to this surface is $\mathbf{n} = -\mathbf{i}_r$

$$\therefore \qquad \sigma_1 = -\mathbf{P}_2\,.\,\mathbf{i}_r\,|_{\,r=b}$$

Now $$\phi_2 = \left(Cr + \frac{D}{r^2}\right)\cos\theta$$

$$= \left[\frac{(2K+1)}{3K}r + \frac{(K-1)b^3}{3K}\frac{1}{r^2}\right]F\cos\theta$$

hence the radial component of the electric field is given by

$$E_{2r} = -\left[(2K+1) - 2(K-1)\frac{b^3}{r^3}\right]\frac{F\cos\theta}{3K}$$

$$\therefore \quad \sigma_1 = (K-1)\epsilon_0\frac{F}{3K}\left[(2K+1) - 2(K-1)\right]\cos\theta$$

$$= (K-1)\frac{\epsilon_0}{K}F\cos\theta$$

Similarly, for the charge distribution over the surface at $r = a$ we have

$$\sigma_2 = -\left[(2K+1) - 2(K-1)\frac{b^3}{a^3}\right]\frac{F\cos\theta}{3K}(K-1)\epsilon_0$$

Now these bound charge distributions σ_1 and σ_2 lie in the radial field of the point charge q and will therefore experience a force–this being the force transmitted to the dielectric shell.

Consider the force on σ_1 (see *Figure 2.5*)

The components of the force radial to the z axis will cancel leaving only the z component.

$$d\mathscr{F}_1 = \left\{\frac{(K-1)}{K}\epsilon_0 F\cos\theta\right\}\{b\ d\theta\ 2\pi b\sin\theta\}\left\{\frac{q}{4\pi\epsilon_0 b^2}\right\}\cos\theta$$

$$\therefore \quad \mathscr{F}_1 = (K-1)\frac{Fq}{2K}\int_0^\pi \cos^2\theta\sin\theta\ d\theta$$

$$\therefore \quad \mathscr{F}_1 = \frac{(K-1)Fq}{3K}$$

Similarly for the other layer of charge at radius a.

$$\mathrm{d}\mathscr{F}_2 = \left\{-(K-1)\epsilon_0\left[(2K+1) - 2(K-1)\frac{b^3}{a^3}\right]\frac{F\cos\theta}{3K}\right\}$$

$$\times\ \{a\,\mathrm{d}\theta\ 2\pi a\sin\theta\}\left\{\frac{q}{4\pi\epsilon_0 a^2}\right\}\cos\theta$$

$$\therefore\quad \mathscr{F}_2 = -(K-1)\left[(2K+1) - 2(K-1)\frac{b^3}{a^3}\right]\frac{Fq}{6K}\int_0^\pi \cos^2\theta\sin\theta\,\mathrm{d}\theta$$

$$\therefore\quad \mathscr{F}_2 = -(K-1)\frac{Fq}{9K}\left[(2K+1) - 2(K-1)\frac{b^3}{a^3}\right]$$

Thus total force on the dielectric, $\mathscr{F}_1 + \mathscr{F}_2$,

$$= -\frac{Fq}{9Ka^3}\left[(2K+1)(K-1)a^3 - 2(K-1)^2b^3 - 3(K-1)a^3\right]$$

$$= -\frac{Fq}{9Ka^3}\left[(2K^2-4K+2)a^3 - 2(K-1)^2b^3\right]$$

$$= -\frac{Fq}{9Ka^3}\left[(2K^2+5K+2)a^3 - 2(K-1)^2b^3 - 9a^3K\right]$$

$$= qE + qF.$$

Finally we see, as we expect, that the total force on the point charge plus dielectric is simply qE.

2.8. Consider

$$\nabla\cdot(K\nabla\phi) = \nabla\phi\cdot\nabla K + K\nabla^2\phi = 0$$

Now
$$\nabla K = a^2\nabla\left(\frac{1}{r^2}\right) = \mathbf{i}_r\left(-\frac{2a^2}{r^3}\right)$$

and $$\nabla \phi = \mathbf{i}_r \frac{\partial \phi}{\partial r} + \mathbf{i}_\theta \frac{1}{r} \frac{\partial \phi}{\partial \theta} + \mathbf{i}_\psi \frac{1}{r \sin \theta} \frac{\partial \phi}{\partial \psi}$$

$$\therefore \quad \nabla \phi . \nabla K = -\frac{2a^2}{r^3} \frac{\partial \phi}{\partial r}$$

$$\therefore \quad \nabla . (K \nabla \phi) = -\frac{2a^2}{r^3} \frac{\partial \phi}{\partial r} + \frac{a^2}{r^2} \left[\frac{1}{r^2} \frac{\partial}{\partial r} \left(r^2 \frac{\partial \phi}{\partial r} \right) + \frac{1}{r^2 \sin \theta} \frac{\partial}{\partial \theta} \left(\sin \theta \frac{\partial \phi}{\partial \theta} \right) \right]$$

dividing throughout by a^2/r^4 and cancelling terms we obtain

$$r^2 \frac{\partial^2 \phi}{\partial r^2} + \frac{1}{\sin \theta} \frac{\partial}{\partial \theta} \left(\sin \theta \frac{\partial \phi}{\partial \theta} \right) = 0$$

Let us now assume a separation of variables solution, $\phi = f(r)\,\Theta(\theta)$. Substituting this into the above equation leads to the separated equations:

$$\frac{1}{\sin \theta} \frac{\mathrm{d}}{\mathrm{d}\theta} \left(\sin \theta \frac{\mathrm{d}\Theta}{\mathrm{d}\theta} \right) + k\Theta = 0$$

$$r^2 \frac{\mathrm{d}^2 f}{\mathrm{d}r^2} = kf$$

k being the separation constant.

The first of these we recognise as Legendre's differential equation, and since our problem includes the polar axis, k is of the form $n(n + 1)$ where n is integer. Putting this into the second equation we see by inspection that this has a solution

$$f(r) = A_n r^{(n+1)} + B_n r^{-n}$$

Thus the complete solution inside the dielectric shell has terms of the form

$$\phi = (A_n r^{(n+1)} + B_n r^{-n})\, P_n$$

Outside of the dielectric the field satisfies $\nabla^2 \phi = 0$, and the potential is of the form $(C_n r^n + D_n r^{-(n+1)})\, P_n$

The boundary conditions to be satisfied are:

(*i*) as $r \to \infty$, $\phi_3 \to -Er\cos\theta$ (see *Figure 2.6*)

(*ii*) at $r = a$, $\phi_2 = \phi_3$

(*iii*) at $r = \lambda a$, $\phi_1 = \phi_2$

(*iv*) at $r = a$, $\dfrac{\partial\phi_3}{\partial r} = K(a)\dfrac{\partial\phi_2}{\partial r} = \dfrac{\partial\phi_2}{\partial r}$

(*v*) at $r = \lambda a$, $\dfrac{\partial\phi_1}{\partial r} = K(\lambda a)\dfrac{\partial\phi_2}{\partial r} = \dfrac{1}{\lambda^2}\dfrac{\partial\phi_2}{\partial r}$

(*vi*) ϕ_1 finite as $r \to 0$.

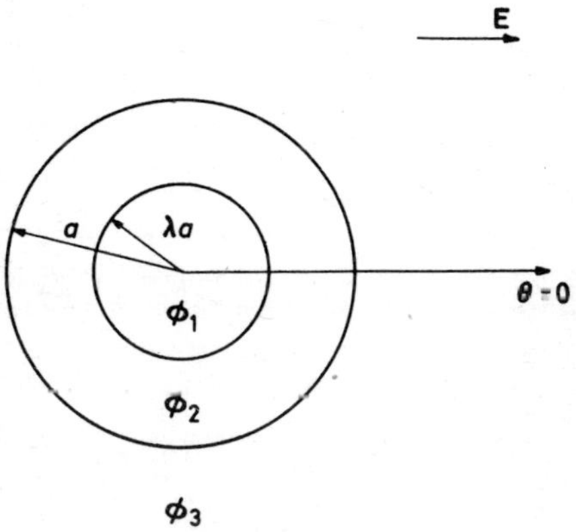

Figure 2.6.

Using (*i*) and (*vi*) we select potentials of the form:

$$\phi_3 = \left(-Er + \frac{A}{r^2}\right)\cos\theta$$

$$\phi_2 = \left(Br^2 + \frac{C}{r}\right)\cos\theta$$

$$\phi_1 = Dr\cos\theta$$

Using (*ii*) $$-Ea + \frac{A}{a^2} = Ba^2 + \frac{C}{a}$$

Using (*iii*) $$D\lambda a = B\lambda^2 a^2 + \frac{C}{\lambda a}$$

Using (*iv*) $$-E - \frac{2A}{a^3} = 2Ba - \frac{C}{a^2}$$

Using (*v*) $$D = \frac{1}{\lambda^2}\left[2B\lambda a - \frac{C}{\lambda^2 a^2}\right]$$

Solving for D in terms of E we get

$$D = \frac{-9E\lambda}{4 + 4\lambda^2 + 2\lambda^3 - \lambda^5}$$

$$A, B, C \neq 0$$

We have satisfied all of the boundary conditions and therefore from the uniqueness theorem we know we have the correct solution. From the form of ϕ_1, we see that the electric field within the cavity is uniform and directed along the polar axis with a magnitude $|D|$.

In the process of solving for the field in the cavity, we have also shown that the potential inside the dielectric is of the form $f(r)\cos\theta$; and since the only value of n required was $n = 1$, it follows from the separated equation that

$$\frac{d^2 f}{dr^2} = \frac{2f}{r^2}$$

2.9. Let us first consider the potential due to a uniformly charged ring. *Figure 2.7* shows the geometry involved.

We know that the field satisfies Laplace's equation at all points other than on the ring itself and must be finite at the origin of the coordinates, and tend to zero as $r \to \infty$. No single potential function can satisfy these

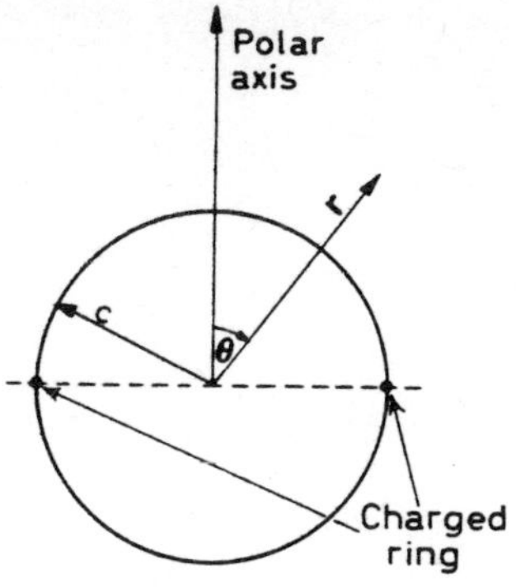

Figure 2.7.

conditions simultaneously so we shall define a potential ϕ_1 to be valid for $r < c$ and ϕ_2 for $r > c$; i.e.

$$\phi_1 = \sum_{n=0}^{\infty} A_n r^n P_n(\mu)$$

$$\phi_2 = \sum_{n=0}^{\infty} B_n r^{-(n+1)} P_n(\mu)$$

For all points on the sphere of radius c, other than at $\theta = \pi/2$, we have $\phi_1 = \phi_2$. This gives

$$B_n = A_n C^{2n+1}$$

To determine the remaining constant, A_n, we use the fact that there will be a discontinuity of the normal derivative of ϕ across the sphere, i.e.

$$-\epsilon_0 \left[\frac{\partial \phi_2}{\partial r} - \frac{\partial \phi_1}{\partial r} \right]_{r=c} = \sigma_s$$

where σ_s is the surface charge density over the sphere. Because of the nature of the charge distribution, it will facilitate matters to define the surface charge density $\sigma_s \delta [\theta - (\pi/2)]$ where δ is the delta function defined by

$$\delta [\theta - (\pi/2)] \equiv 0, \theta \neq (\pi/2)$$

$$\equiv 1, \theta = (\pi/2)$$

and $$\int_0^{\pi} (c\, d\theta\, 2\pi c \sin\theta)\, \sigma_s\, \delta [\theta - (\pi/2)] = Q$$

i.e. $$\sigma_s = \frac{Q}{2\pi c^2}$$

$$\therefore \quad \sum_{n=0}^{\infty} -\epsilon_0 \left[(-1)(n+1) r^{-(n+2)} c^{2n+1} - n r^{n-1} \right]_{r=c} A_n P_n(\mu)$$

$$= \sigma_s\, \delta [\theta - (\pi/2)]$$

i.e. $$\sum_{n=0}^{\infty} (2n+1) c^{n-1} A_n P_n(\mu) = \frac{\sigma_s}{\epsilon_0} \delta [\theta - (\pi/2)]$$

Now to find the A_n we multiply both sides of this equation by $P_n(\mu)$ and integrate over the range of μ (cos θ), -1 to $+1$. Making use of the integral

$$\int_{-1}^{+1} P_n(\mu)\, P_m(\mu)\, d\mu = \frac{2}{(2n+1)} \delta_{mn}$$

this gives

$$A_n = \frac{Q\, P_n(0)}{4\pi\epsilon_0 c^{n+1}}$$

Hence $$\phi_1 = \frac{Q}{4\pi\epsilon_0 c} \sum_{n=0}^{\infty} \left(\frac{r}{c}\right)^n P_n(0) P_n(\mu)$$

and
$$\phi_2 = \frac{Q}{4\pi\epsilon_0 c} \sum_{n=0}^{\infty} \left(\frac{c}{r}\right)^{(n+1)} P_n(0)P_n(\mu).$$

Let us now turn our attention to finding the charge on the inner of the two earthed spheres (see *Figure 2.8*).

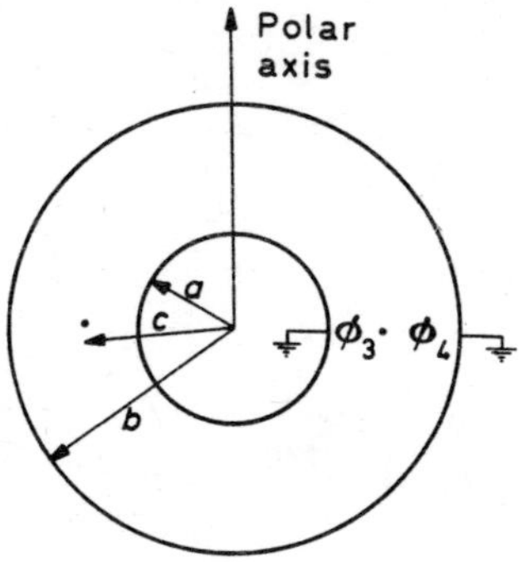

Figure 2.8.

The potential without the charged ring would be of the form

$$\sum_{n=0}^{\infty} (D_n r^n + F_n r^{-(n+1)})P_n(\mu)$$

and onto this we superpose the potential due to the ring, giving

$$\phi_3 = k \sum_{n=0}^{\infty} \left(\frac{r}{c}\right)^n P_n(0)P_n(\mu) + \sum_{n=0}^{\infty} (D_n r^n + F_n r^{-(n+1)})P_n(\mu)$$

$$k = \left(\frac{Q}{4\pi\epsilon_0 c}\right)$$

$$\phi_4 = k \sum_{n=0}^{\infty} \left(\frac{c}{r}\right)^{(n+1)} P_n(0)P_n(\mu) + \sum_{n=0}^{\infty} (D_n r^n + F_n r^{-(n+1)})P_n(\mu)$$

These potentials are only restricted by the conditions that $\phi_3 = 0$ at $r = a$, and $\phi_4 = 0$ at $r = b$.

Applying these and equating the coefficients of $P_n(\mu)$ in each case gives us two equations involving D_n and F_n, i.e.

$$0 = k\left(\frac{a}{c}\right)^n P_n(0) + D_n a^n + \frac{F_n}{a^{n+1}}$$

$$0 = k\left(\frac{c}{b}\right)^{n+1} P_n(0) + D_n b^n + \frac{F_n}{b^{n+1}}$$

Solving for D_n and F_n

$$D_n = \frac{kP_n(0)}{c^n}\left[\frac{c^{2n+1} - a^{2n+1}}{a^{2n+1} - b^{2n+1}}\right]$$

$$F_n = \frac{kP_n(0)}{c^n}\left[\frac{b^{2n+1} - c^{2n+1}}{a^{2n+1} - b^{2n+1}}\right] a^{2n+1}$$

The charge density on the inner sphere is simply

$$\sigma_s = -\epsilon_0 \frac{\partial \phi_3}{\partial r}\bigg|_{r=a} = -k\epsilon_0 \sum_{n=0}^{\infty} \frac{P_n(0)P_n(\mu)}{c^n} a^{n-1}$$

$$\times \left\{n + \left(\frac{c^{2n+1} - a^{2n+1}}{a^{2n+1} - b^{2n+1}}\right) n - \left(\frac{b^{2n+1} - c^{2n+1}}{a^{2n+1} - b^{2n+1}}\right)(n+1)\right\}$$

$$= \frac{Q}{4\pi c} \sum_{n=0}^{\infty} \frac{a^{n-1}}{c^n}(2n+1)\left(\frac{c^{2n+1} - b^{2n+1}}{b^{2n+1} - a^{2n+1}}\right) P_n(0)P_n(\mu)$$

Hence charge on the inner conductor, Q_1, is obtained by integrating σ_s over the inner sphere, i.e.

$$Q_1 = \int_0^{\pi} \sigma_s(2\pi a \sin\theta)a \, d\theta$$

$$= \frac{Qa}{2c} \sum_{n=0}^{\infty} \left(\frac{a}{c}\right)^n (2n+1)\left(\frac{c^{2n+1} - b^{2n+1}}{b^{2n+1} - a^{2n+1}}\right) P_n(0) \int_0^{\pi} P_n(\mu) \sin\theta \, d\theta$$

We can see from the symmetry of the problem that the induced charge distribution on the inner sphere must have even symmetry about $\theta = \pi/2$. It follows then that n cannot be odd. Now for all n even

$$\int_0^{\pi} P_n(\mu) \sin\theta \, d\theta = 0$$

$$\therefore \qquad Q_1 = -Q\frac{a}{c} \cdot \left(\frac{b-c}{b-a}\right)$$

2.10. In this problem we must define two potential functions, ϕ_1 and ϕ_2, one valid for the range $0 \leqslant r \leqslant a$ and the other for $a \leqslant r < c$. A suitable choice of potentials is

$$\phi_1 = \sum_{n=0}^{\infty} A_n r^n P_n(\mu)$$

$$\phi_2 = \sum_{n=0}^{\infty} \left\{\frac{B_n}{r^{n+1}} + \frac{\alpha}{K}\left(\frac{r}{c}\right)^n P_n(0)\right\} P_n(\mu)$$

where α is $e/4\pi\epsilon_0 c$ and we have used the expansion for the potential of the ring of charge we proved earlier in Problem 2.9.

These potentials are subject to the boundary conditions that at $r = a$, $\phi_1 = \phi_2$, and $\partial\phi_1/\partial r = K(\partial\phi_2/\partial r)$.

Applying these conditions to the above potentials and equating coefficients of $P_n(\mu)$ on each side of the resulting equations gives us two equations involving the unknown constants, i.e.

$$a^n A_n = \frac{\alpha}{K}\frac{a^n}{c^n} P_n(0) + \frac{B_n}{a^{n+1}}$$

$$na^{n-1} A_n = \alpha \frac{na^{n-1}}{c^n} P_n(0) - \frac{(n+1)B_n K}{a^{n+2}}$$

Eliminating B_n from the above equations gives

$$A_n = \frac{\alpha P_n(0)}{c^n} \left\{ \frac{2n + 1}{K(n + 1) + n} \right\}$$

$$\therefore \quad \phi_1 = \alpha \sum_{n=0}^{\infty} \left(\frac{r}{c}\right)^n P_n(0) \left\{ \frac{2n + 1}{K(n + 1) + n} \right\} P_n(\mu)$$

We know, from Rodrigue's formula, that

$$P_n(\mu) = \frac{(-1)^n}{2^n\, n!} \frac{\mathrm{d}^n}{\mathrm{d}\mu^n} (1 - \mu^2)^n$$

Expanding $(1 - \mu^2)^n$ gives

$$1 - n\mu^2 + \frac{n(n-1)}{2!}\mu^4 - \frac{n(n-1)(n-2)\mu^6}{3!} + \ldots + \mu^{2n}$$

Thus when we differentiate this n times and then put $\mu = 0$, the only term remaining will be the one involving μ^n, provided that n is even, and none will remain if it is not. Hence

$$P_n(0) = \frac{1}{2^n n!} \left[(-1)^{n/2} \frac{n!\, n!}{[(n/2)!]^2} \right] \qquad (n \text{ even})$$

$$\therefore \quad \phi_1 = \alpha \sum_{\text{all even } n} \left(\frac{r}{c}\right)^n \frac{(-1)^{n/2}\, n!}{2^n [(n/2)!]^2} \left\{ \frac{(2n + 1)}{K(n + 1) + n} \right\} P_n(\mu)$$

$$\therefore \quad \phi_1 = \frac{e}{4\pi\epsilon_0 c} \sum_{n=0}^{\infty} \frac{4n + 1}{K(2n + 1) + 2n} \; \frac{(-1)^n (2n)!}{2^{2n} (n!)^2} \left(\frac{r}{c}\right)^{2n} P_{2n}(\mu)$$

2.11. Following the arguments of Problem 2.9 we can immediately write down the potential due to the two hoops:

$$\phi_1 = \frac{e}{4\pi\epsilon_0\sqrt{2}\,b}\sum_{n=0}^{\infty}\left(\frac{r}{\sqrt{2}\,b}\right)^n P_n(\mu)\left\{P_n\left(-\frac{1}{\sqrt{2}}\right) - P_n\left(\frac{1}{\sqrt{2}}\right)\right\}$$

where we have replaced c, the distance from the origin to the ring of charge, by $\sqrt{2}\ b$ and the angles in the delta function are now $\pi/4$ and $3\pi/4$.

Inserting the first few terms of $P_n\ [\pm(1/\sqrt{2})]$ leads to

$$\phi_1 = \frac{e}{4\pi\epsilon_0\sqrt{2}\,b}\left[-\frac{r}{b}P_1(\mu) + \frac{r^3}{8b^3}P_3(\mu) + \ldots\right]$$

Now the potential set up by the induced charges on the earthed sphere can be written as

$$\phi_2 = \sum_{n=0}^{\infty} A_n\, r^{-(n+1)} P_n(\mu)$$

and we know that the total potential, $\phi_1 + \phi_2$, over the surface of the sphere must be zero; i.e.

$$\left(\frac{A_1}{a^2} - \frac{e}{4\pi\epsilon_0\sqrt{2}}\frac{a}{b^2}\right)P_1(\mu) + \left(\frac{A_3}{a^4} + \frac{e}{4\pi\epsilon_0\sqrt{2}}\frac{1}{8}\frac{a^3}{b^4}\right)P_3(\mu) + \ldots = 0$$

$$\therefore \qquad \phi_2 = \frac{e}{4\pi\epsilon_0\sqrt{2}}\frac{a^3}{b^2}\left[\frac{P_1(\mu)}{r^2} - \frac{a^4 P_3(\mu)}{8b^2r^4} + \ldots\right]$$

The electric field set up at $r = b$ due to the ring at b is clearly zero. By symmetry, the field at b due to the induced charges on the sphere and the other ring both act along the positive z axis, hence the increase in the field due to the sphere is simply $-\partial\phi_2/\partial r$ evaluated at $r = b$; i.e.

$$\Delta E = \frac{e\sqrt{2}\,a^3}{4\pi\epsilon_0\, b^5}\left[1 + 0\left(\frac{a^4}{b^4}\right)\right]$$

2.12. The potential inside the spherical cavity will have a contribution from the point charge and a further contribution from the bound polarization charge in the dielectric medium. The first of these can be expanded in terms of Legendre polynomials about the centre of the sphere as origin:

$$\frac{p}{d}\sum\left(\frac{r}{d}\right)^n P_n(\mu) \quad \text{for} \quad r<d$$

$$\left(p=\frac{e}{4\pi\epsilon_0}\right)$$

$$\frac{p}{r}\sum\left(\frac{d}{r}\right)^n P_n(\mu) \quad \text{for} \quad r>d$$

and the second is simply

$$\sum C_n r^n P_n(\mu)$$

Hence the total potential inside the cavity for $d<r<a$ is

$$\phi_1=\sum\left(\frac{pd^n}{r^{n+1}}+C_n r^n\right)P_n(\mu)$$

Outside the cavity the potential is

$$\phi_2=\sum\frac{B_n P_n(\mu)}{r^{n+1}}$$

The boundary conditions to be satisfied are that at $r=a$, $\phi_1=\phi_2$ and

$$\frac{\partial\phi_1}{\partial r}=\epsilon\frac{\partial\phi_2}{\partial r}$$

Applying these conditions gives us

$$\frac{B_n}{a^{n+1}}=\frac{pd^n}{a^{n+1}}+C_n a^n$$

$$\frac{-\epsilon B_n(n+1)}{a^{n+2}}=-\frac{pd^n(n+1)}{a^{n+2}}+C_n n a^{n-1}$$

which solving for C_n yields

$$C_n = -\frac{(\epsilon - 1)pd^n}{a^{2n+1}}\left\{\frac{n+1}{\epsilon(n+1)+n}\right\}$$

The potential within the cavity is then:

$$\phi_1 = \sum pd^n\left(\frac{1}{r^{n+1}} - \frac{(\epsilon - 1)(n+1)r^n}{a^{2n+1}\{\epsilon(n+1)+n\}}\right)P_n(\mu) \qquad (r > d)$$

$$\phi_1 = \sum pr^n\left(\frac{1}{d^{n+1}} - \frac{(\epsilon - 1)(n+1)d^n}{a^{2n+1}\{\epsilon(n+1)+n\}}\right)P_n(\mu) \qquad (r < d)$$

Now the electric field within the cavity due to the bound charges is

$$-\frac{\partial}{\partial r}\left[\sum C_n r^n\right]_{r=d}$$

at the position of the point charge and is directed away from the centre.

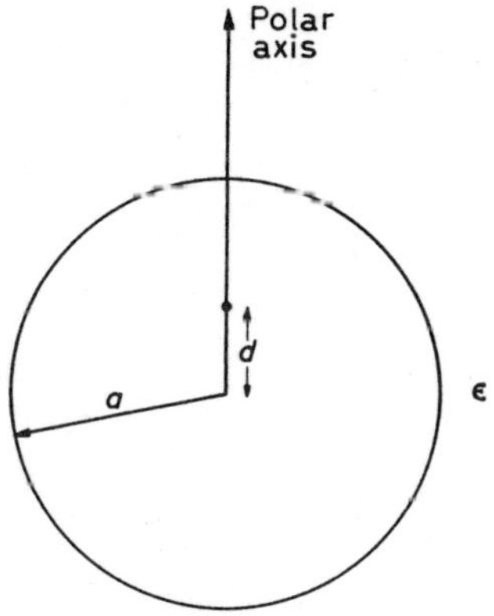

Figure 2.9.

We have here made use of the symmetry of the problem (see *Figure 2.9*) and also that $P_n(1) = 1$.

Hence the force on the point charge e is

$$p\frac{(\epsilon - 1)e}{ad}\sum\frac{n(n+1)}{\epsilon(n+1)+n}\left(\frac{d}{a}\right)^{2n}$$

2.13. We wish to determine the magnetic field over the loop of radius b set up by the larger loop of radius a carrying the current I. To do this we shall solve for the scalar magnetic potential in spherical polar coordinates for the region $r < a$.

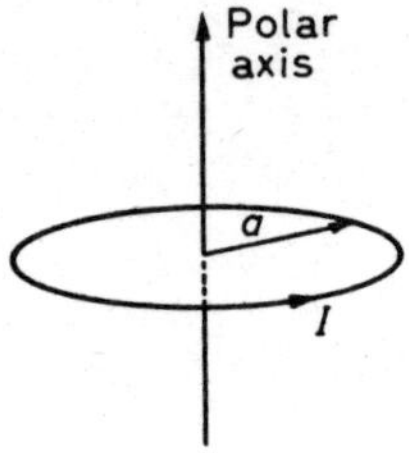

Figure 2.10.

Now the potential satisfies Laplace's equation and must be finite at the origin (see *Figure 2.10*). Hence

$$\phi = \sum A_n r^n P_n(\mu)$$

The coefficients A_n can be obtained by a rather indirect approach, the basis of the method being simply that it is an easy exercise to find the potential along the polar axis. This is

$$\phi' = \frac{I}{4\pi}\Omega$$

where Ω is the solid angle subtended by the current loop at the field point, i.e.

$$\phi' = \frac{I}{2}\left(1 - \frac{r}{\sqrt{r^2 + a^2}}\right)$$

and we can expand this as a power series in (r/a):

$$\phi' = \frac{I}{2}\left[1 - \frac{r}{a}\left\{1 + \frac{r^2}{a^2}\right\}^{-1/2}\right]$$

$$\therefore \quad \phi' = \frac{I}{2}\left[1 - \frac{r}{a} + \frac{1}{2}\frac{r^3}{a^3} - \frac{3}{8}\frac{r^5}{a^5} + \ldots\right] \qquad \ldots(1)$$

But $\phi' = \sum A_n r^n$ since all $P_n(1) = 1$ and the A_n may be obtained from Equation (1). Hence

$$\phi = \frac{I}{2}\left[1 - \frac{r}{a}P_1 + \frac{1}{2}\frac{r^3}{a^3}P_3 - \frac{3}{8}\frac{r^5}{a^5}P_5 + \ldots\right]$$

Now the force on an element of the smaller loop, $b\ \mathrm{d}\theta$, carrying a current I' is $B_r(\theta)\ b\ \mathrm{d}\theta\ I'$ and the radial magnetic field over the loop b is

$$B_r(\theta) = -\mu_0 \left.\frac{\partial \phi}{\partial r}\right|_{r=b}$$

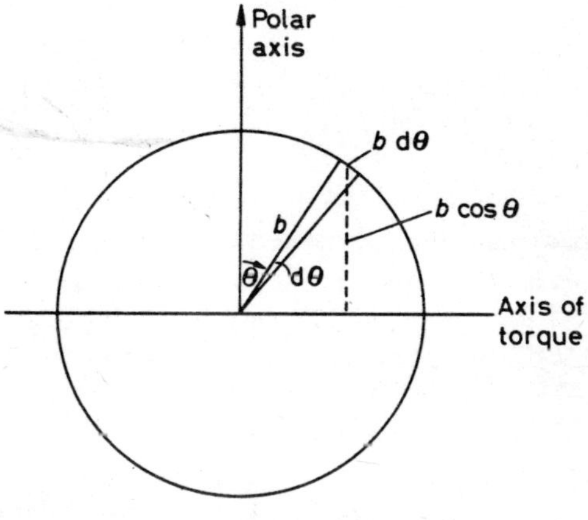

Figure 2.11.

Thus by symmetry (see *Figure 2.11*), total torque on the loop is

$$T = 4I'b^2 \int_0^{\pi/2} B_r(\theta)\cos\theta\ \mathrm{d}\theta$$

$$= \mu_0 4I'b^2 \int_0^{\pi/2} \left(\frac{I}{2a}\right)\left(P_1 - \frac{3}{2}\frac{b^2}{a^2}P_3 + \ldots\right)\cos\theta\ \mathrm{d}\theta$$

But
$$\int_0^{\pi/2} P_1 \cos\theta \, d\theta = \int_0^{\pi/2} \cos^2\theta \, d\theta = \frac{\pi}{4}$$

and
$$-\frac{3}{2}\frac{b^2}{a^2}\int_0^{\pi/2} P_3 \cos\theta \, d\theta = -\frac{3}{2}\frac{b^2}{a^2}\frac{1}{2}\int_0^{\pi/2} (5\cos^4\theta - 3\cos^2\theta) \, d\theta$$

$$= -\frac{9}{16}\frac{b^2}{a^2}\frac{\pi}{4}$$

The next term in the series is of order (b^4/a^4),

$$\therefore \qquad T = \frac{\mu_0 \pi}{2a} I I' b^2 \left[1 - \frac{9}{16}\frac{b^2}{a^2} + 0\left(\frac{b^4}{a^4}\right)\right]$$

2.14. Let us determine the field due to the current sheet at $y = 0$. We will do this by first determining the scalar magnetic potential in the region above the current sheet by solving Laplace's equation assuming a separation of variables solution.

Assuming $\phi = X(x)\,Y(y)$ gives the separated equations

$$\frac{1}{X}\frac{d^2X}{dx^2} = -\frac{1}{Y}\frac{d^2Y}{dy^2} = \text{a constant}$$

Now we are looking for a solution which is periodic in the x direction and has the form

$$X = A\cos\left(\frac{2\pi}{L}x\right) + B\sin\left(\frac{2\pi}{L}x\right)$$

hence

$$\phi = \left\{A\cos\left(\frac{2\pi}{L}x\right) + B\sin\left(\frac{2\pi}{L}x\right)\right\}\{C\,e^{-(2\pi/L)y} + D\,e^{(2\pi/L)y}\}$$

We now apply the boundary condition that the difference in tangential H across the $y = 0$ plane must be the current density on this plane which is given; thus, using symmetry,

$$-\left(\frac{2\pi}{L}\right)\left[B\cos\left(\frac{2\pi}{L}x\right) - A\sin\left(\frac{2\pi}{L}x\right)\right][C + D] = \frac{K_1}{2}\cos\left(\frac{2\pi x}{L}\right)$$

$$\therefore \qquad A = 0 \quad \text{and} \quad B(C + D) = -\frac{K_1 L}{4\pi}$$

We strongly suspect that the potential should tend to zero as $y \to \infty$ and this can be confirmed quite easily by looking at H on the y axis (see *Figure 2.12*).

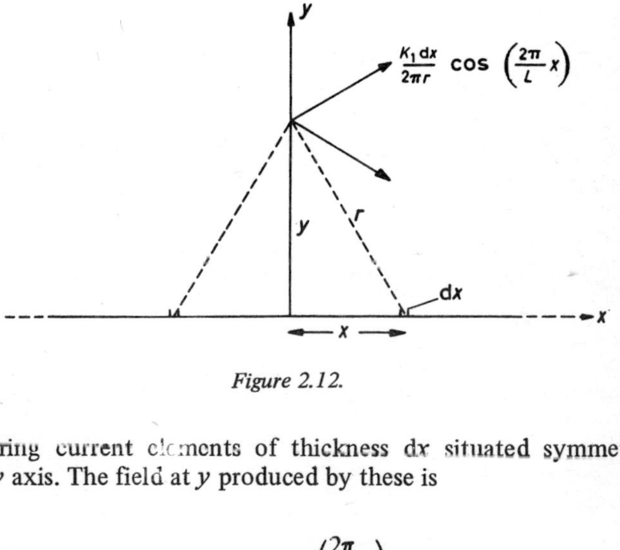

Figure 2.12.

Considering current elements of thickness dx situated symmetrically about the y axis. The field at y produced by these is

$$dH = 2 \cdot \frac{y}{r} \cdot \frac{K_1 \cos\left(\frac{2\pi}{L}x\right) dx}{2\pi r}$$

$$\therefore \qquad H = K_1 \frac{y}{\pi} \int_0^{\infty} \frac{\cos\left(\frac{2\pi}{L}x\right) dx}{(y^2 + x^2)}$$

The integral can be done using contour methods and the result is

$$H = \frac{K_1}{2} e^{-(2\pi/L)y}$$

This is the form we expected. However we need not bother putting ourselves to the effort, all we need note is that

$$H = K_1 \frac{y}{\pi} \int_0^\infty \frac{\cos[(2\pi/L)x]\,dx}{y^2 + x^2} < K_1 \frac{y}{\pi} \int_0^\infty \frac{dx}{y^2 + x^2} = \frac{K_1}{2}$$

Clearly then an exponentially increasing term is not possible, thus $D = 0$. Hence

$$\phi = -\frac{K_1 L}{4\pi} \sin\left(\frac{2\pi}{L}x\right) e^{-(2\pi/L)y}$$

$$\therefore \quad B_x = -\mu_0 \frac{\partial \phi}{\partial x} = \mu_0 \frac{K_1}{2} \cos\left(\frac{2\pi}{L}x\right) e^{-(2\pi/L)y}$$

$$B_y = -\mu_0 \frac{\partial \phi}{\partial y} = -\mu_0 \frac{K_1}{2} \sin\left(\frac{2\pi}{L}x\right) e^{-(2\pi/L)y}$$

If F_x and F_y are the x and y components of the force per periodic length L, and per unit length in the z direction, then

$$F_x = \int_0^L B_y(x, y = d)\,(K_z'\,dx)$$

$$= -\frac{\mu_0 K_1 K_2}{2} \cdot e^{-(2\pi/L)d} \int_0^L \sin\left(\frac{2\pi}{L}x\right) \sin\left(\frac{2\pi x}{L} + \frac{2\pi l}{L}\right) dx$$

$$\therefore \quad F_x = -\frac{\mu_0 K_1 K_2\, e^{-(2\pi/L)d}\, L}{4} \cdot \cos\left(\frac{2\pi l}{L}\right)$$

$$F_y = \int_0^L -B_x(x, y = d)\,(K_z'\,\mathrm{d}x)$$

$$= -\frac{\mu_0 K_1 K_2\, \mathrm{e}^{-(2\pi/L)d}}{2} \int_0^L \cos\left(\frac{2\pi}{L}x\right) \sin\left(\frac{2\pi x}{L} + \frac{2\pi l}{L}\right) \mathrm{d}x$$

$$\therefore \qquad \underline{F_y = -\frac{\mu_0 K_1 K_2\, \mathrm{e}^{-(2\pi/L)d}\, L}{4} \sin\left(\frac{2\pi l}{L}\right)}$$

We see from these expressions that the force vector has a magnitude independent of l and that it makes an angle of $(2\pi l/L)$ with the $y = 0$ plane.

CHAPTER 3

3.1. The geometrical arrangement of this problem is shown in *Figure 3.7.* Since the cross-section of the system is independent of the third coordinate it is clear that we are concerned with a two-dimensional problem which should be amenable to the method of conformal transformation. The first step in such a solution is to decide on the best (in the sense of the most convenient) configuration to start from on the complex z-plane. Because of possible symmetry in the physical system it is not always necessary to consider the whole system in a solution by this method. As this is the first problem of the present chapter we will consider the possible reductions at some length.

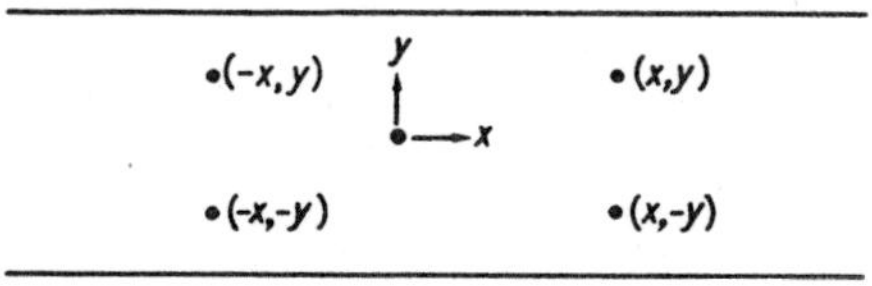

Figure 3.7.

First of all we observe that the two-dimensional structure is symmetrical about both the x and y axes. In effect, this means that the potentials at the four points (x, y), $(-x, y)$, $(x, -y)$ and $(-x, -y)$ are identical. Since

$$\mathbf{E} = -\nabla\phi$$

we can easily show that, in this case $E_x = 0$ when $x = 0$ and that $E_y = 0$ when $y = 0$ or, physically, that the flow lines of the electric field never cross either the x-axis or the y-axis. These axes are in fact two of the flow lines. We will call surfaces which contain the flow lines *magnetic walls* and denote them by a heavy dashed line and we shall call surfaces which are perpendicular to the flow lines *electric walls* and denote them by a solid line. (These are the same as the equipotential surfaces in actuality, although in the solution of problems only those surfaces which simplify the solution are designated as electric walls, and in the present problem they will not be used except insofar as the conductors themselves are electric walls.)

As a result of the above discussion the physical configuration of *Figure 3.7* contains two magnetic walls as shown in *Figure 3.8.*

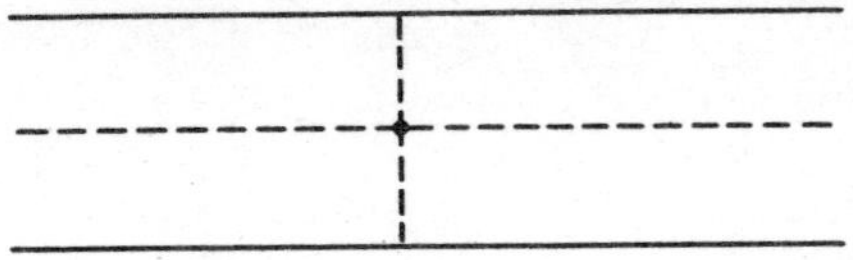

Figure 3.8.

We are now faced with four alternatives:

(*a*) we can place the whole configuration on the complex z-plane, *or*

(*b*) we can place the right hand half (or, of course, the left hand half) on the complex z-plane, *or*

(*c*) we can place the upper half (or lower half) on the complex z-plane, *or*

(*d*) we can place one quadrant on the complex z-plane.

In making the final decision as to which of these alternatives to choose we must consider the type of transformation that we propose to employ. Assuming that a dictionary of transformations is not available the contours of any of the alternatives can be transformed to the real axis of a complex w-plane by the Schwarz-Christoffel transformation. We can in this event deduce the resultant w-plane configurations for the four cases and these are shown in *Figure 3.9.* In this figure the various points corresponding to the corners of the polygons have been labelled.

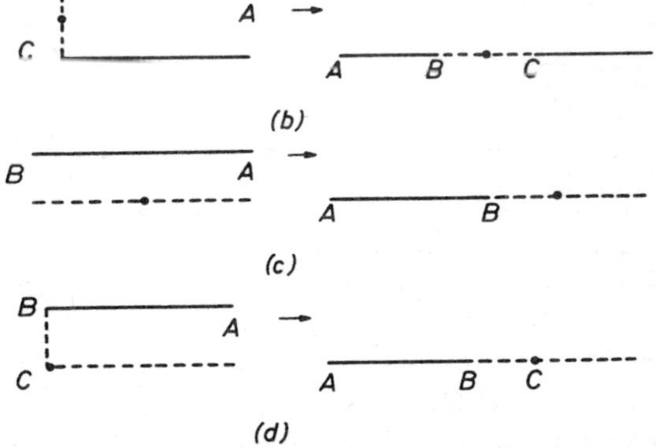

Figure 3.9.

We now consider the relative simplicity of the w-plane configurations. The first alternative has produced a line charge above a grounded plane—a problem easily solved by the method of images. The second does not look particularly promising. The third and fourth alternatives produce identical problems and although we may recognise that a second transformation applied to these (a square root law transformation, in fact) will reduce them to the configuration of *Figure 3.9(a)* the trouble would seem to be hardly worth the effort involved. Thus on grounds of convenience (and it must be realised that this is only algebraic convenience in the case of the first, third and fourth alternatives) we are led to transform the whole system for this particular problem. Later problems will lead to other conclusions.

We are now in a position to start the solution. Since we can place the configuration anywhere on the complex z-plane we shall choose the lines $z = \pm\, jb/2$ to be the conductors and the point $z = 0$ to be the line charge as shown in *Figure 3.10.* As far as the w-plane representation is concerned we

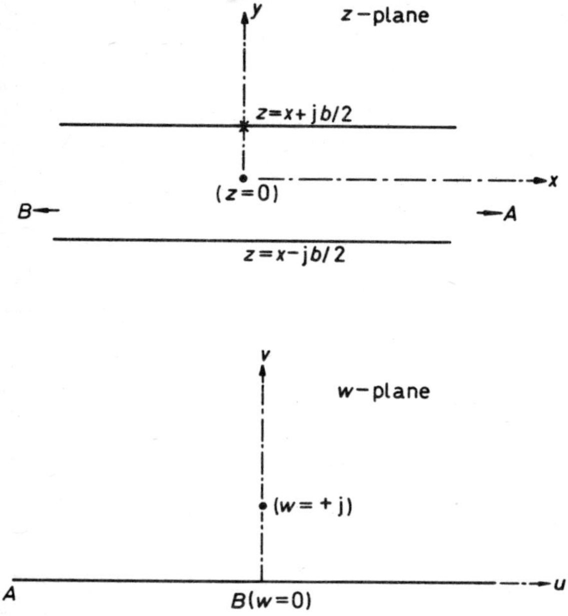

Figure 3.10.

shall choose the point correspondence as shown in this figure so that the lower conductor transforms into the positive half of the real axis and the upper conductor into the negative half. Since the A-vertex on the z-plane goes to infinity on the w-plane we have to consider only the B-vertex which has an included angle of 0°.

The Schwarz-Christoffel differential equation is

$$\frac{dz}{dw} = Pw^{-1}$$

with the solution

$$z = P \ln w + Q \qquad \dots (1)$$

where P and Q are constants to be determined as follows:

If $z = x + jb/2$ then $w = -r = r \exp(j\pi)$ and if $z = x' - jb/2$ then $w = r'$ with r and r' real and positive.

Thus we obtain the two equations

$$\left.\begin{aligned} x + j\frac{b}{2} &= P \ln r + j\pi P + Q \\ x' - j\frac{b}{2} &= P \ln r' + Q \end{aligned}\right\}$$

with solutions

$$P = b/\pi \quad \text{and} \quad Q = -jb/2$$

If these values are inserted into Equation (1) the transformation can be shown to be

$$w = j \exp(\pi z/b) \qquad \dots (2)$$

and since $w = u + jv$ and $z = x + jy$ we obtain

$$\left.\begin{aligned} u &= -\exp\left(\frac{\pi x}{b}\right)\sin\left(\frac{\pi y}{b}\right) \\ v &= \exp\left(\frac{\pi x}{b}\right)\cos\left(\frac{\pi y}{b}\right) \end{aligned}\right\} \qquad \dots (3)$$

It now remains to determine the potential ϕ as a function of the w-plane coordinates u, v, and to substitute for these coordinates the functions of

x, y, of Equation (3). The potential problem in the w-plane is simple and leads to

$$\phi = \frac{\sigma}{4\pi\epsilon} \ln \left\{ \frac{u^2 + (v+1)^2}{u^2 + (v-1)^2} \right\}$$

since the line charge is at $w = \mathrm{j}$.

Substituting from Equation (3) gives the potential function of the original problem in the form

$$\phi = \frac{\sigma}{4\pi\epsilon} \ln \left\{ \frac{\cosh\left(\frac{\pi x}{b}\right) + \cos\left(\frac{\pi y}{b}\right)}{\cosh\left(\frac{\pi x}{b}\right) - \cos\left(\frac{\pi y}{b}\right)} \right\}$$

which can be expressed alternatively as

$$\phi = \frac{\sigma}{2\pi\epsilon} \tanh^{-1} \left\{ \frac{\cos\left(\frac{\pi y}{b}\right)}{\cosh\left(\frac{\pi x}{b}\right)} \right\}$$

The form of the equipotential surfaces is most conveniently determined from the second of the above expressions since, if ϕ is a constant, the equation of the equipotential surface is of the form

$$\cos\left(\frac{\pi y}{b}\right) = K \cosh\left(\frac{\pi x}{b}\right)$$

where

$$K = \tanh\left(\frac{2\pi\epsilon\phi}{\sigma}\right)$$

Very close to the charge, the curves represented by the above equation are circular, a little further out they are almost elliptical and then become rather like flattened ellipses.

3.2. To solve this problem we can use the same transformation as that of Problem 3.1, the essential difference being that the line charge in the z-plane is no longer at the origin but lies at a point on the imaginary y-axis as shown in *Figure 3.11.* This figure also shows the transformed point on the w-plane which is a point on the unit circle.

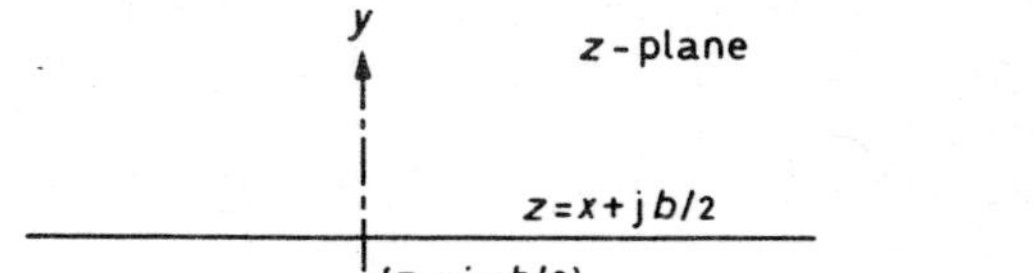

Figure 3.11.

The new potential function is therefore

$$\phi = \frac{\sigma}{4\pi\epsilon} \ln \left\{ \frac{\left[u + \sin\left(\frac{\pi\alpha}{2}\right)\right]^2 + \left[v + \cos\left(\frac{\pi\alpha}{2}\right)\right]^2}{\left[u + \sin\left(\frac{\pi\alpha}{2}\right)\right]^2 + \left[v - \cos\left(\frac{\pi\alpha}{2}\right)\right]^2} \right\}$$

whence, in the z-plane

$$\phi = \frac{\sigma}{4\pi\epsilon} \ln \left\{ \frac{\cosh\left(\frac{\pi x}{b}\right) + \cos\left(\frac{\pi y}{b} + \frac{\pi\alpha}{2}\right)}{\cosh\left(\frac{\pi x}{b}\right) - \cos\left(\frac{\pi y}{b} - \frac{\pi\alpha}{2}\right)} \right\}$$

The potential function for the second part of the problem can be obtained by superposition as

$$\phi = \frac{\sigma}{4\pi\epsilon} \ln \left\{ \frac{\left[\cosh\left(\frac{\pi x}{b}\right) + \cos\left(\frac{\pi y}{b} + \frac{\pi\alpha}{2}\right)\right]\left[\cosh\left(\frac{\pi x}{b}\right) + \cos\left(\frac{\pi y}{b} - \frac{\pi\alpha}{2}\right)\right]}{\left[\cosh\left(\frac{\pi x}{b}\right) - \cos\left(\frac{\pi y}{b} - \frac{\pi\alpha}{2}\right)\right]\left[\cosh\left(\frac{\pi x}{b}\right) - \cos\left(\frac{\pi y}{b} + \frac{\pi\alpha}{2}\right)\right]} \right\}$$

3.3. By an argument based on symmetry similar to that of the first problem it is clear that the gap in the plane of the septum is a magnetic wall and we need to consider only one half of the structure as shown in *Figure 3.12.* This shape is suitable for the application of a Schwarz-Christoffel transformation which will produce the *w*-plane configuration also shown in the figure.

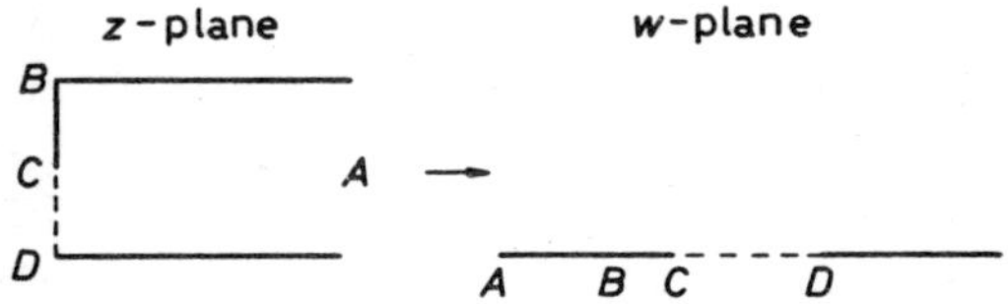

Figure 3.12.

The *w*-plane representation is not immediately soluble, however, and when this happens it is necessary to consider the possibility of a further transformation. Thus we wish to find a soluble configuration on a third complex plane, say the ζ-plane, which can be transformed into the *same* *w*-plane configuration.

In order to formalize this argument let us suppose that the z/w transformation is of the form

$$w = f(z)$$

and that we can find a ζ/w transformation of the form

$$w = g(\zeta)$$

The transformation obtained by eliminating w from these two equations, that is,

$$f(z) = g(\zeta)$$

will transform the z-plane configuration into an immediately soluble ζ-plane configuration.

Now in this problem we do not have to look far for the second transformation for the z-plane configuration when a becomes equal to b is clearly a parallel plate system which is immediately soluble and would therefore be suitable for the ζ-plane configuration. The sequence of transformations required is shown in *Figure 3.13* in which the point correspondences are

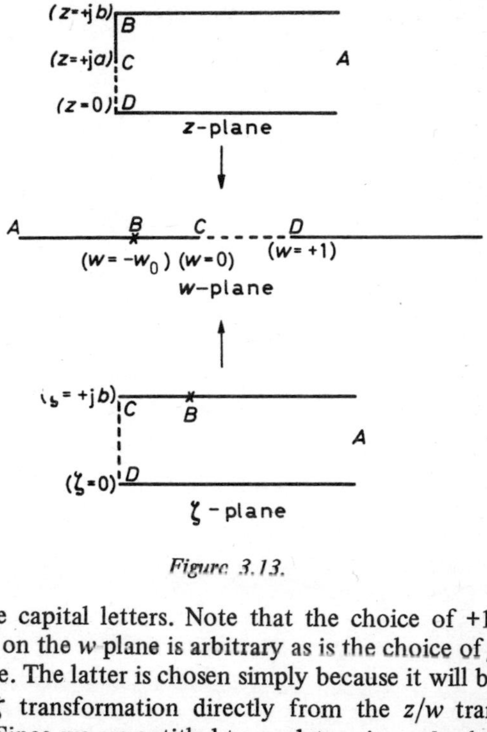

Figure 3.13.

denoted by the capital letters. Note that the choice of +1 and 0 for the points D and C on the w plane is arbitrary as is the choice of jb for the point C on the ζ-plane. The latter is chosen simply because it will be convenient to obtain the w/ζ transformation directly from the z/w transformation by putting $a = b$. Since we are entitled to predetermine only three points in the w-plane (A, C and D) the value of w_0 must be obtained from the transformation.

We are now in a position to start the analysis. The differential equation for the z/w transformation is

$$\frac{dz}{dw} = \frac{P}{\sqrt{\{(w + w_0)(w - 1)\}}}$$

$$= \frac{P}{\sqrt{\left\{\left(w + \frac{w_0 - 1}{2}\right)^2 - \left(\frac{w_0 + 1}{2}\right)^2\right\}}}$$

and, transforming variables by

$$w + \frac{w_0 - 1}{2} = \frac{w_0 + 1}{2} \cosh \theta$$

we obtain

$$z = P\theta + Q$$

We must now determine the constants P, Q and w_0 and this is done from the point correspondence. Note that at this stage it is not usually convenient to eliminate θ before determining the constants and we therefore construct as follows a table of points including some values of θ.

z	0	ja	jb
w	$+1$	0	$-w_0$
θ	0		$j\pi$

Now, since $z = 0$ when $\theta = 0$ then $Q = 0$.
Also, since $z = jb$ when $\theta = j\pi$ then $P = b/\pi$.
Thus

$$\theta = \left(\frac{\pi z}{b}\right)$$

and

$$w + \frac{w_0 - 1}{2} = \frac{w_0 + 1}{2} \cosh \left(\frac{\pi z}{b}\right) \qquad \ldots.(1)$$

Since $w = 0$ when $z = ja$ then $(w_0 - 1) = (w_0 + 1) \cos (\pi a/b)$, whence

$$w_0 = \frac{1 + \cos\left(\frac{\pi a}{b}\right)}{1 - \cos\left(\frac{\pi a}{b}\right)} = \cot^2\left(\frac{\pi a}{2b}\right)$$

If this is inserted into Equation (1) we obtain

$$w = \frac{1}{2}\left[\cosh\frac{\pi z}{b} + 1\right] + \frac{1}{2}\left[\cosh\frac{\pi z}{b} - 1\right]\cot^2\left(\frac{\pi a}{2b}\right)$$

$$= 1 + \sinh^2\left(\frac{\pi z}{2b}\right)\operatorname{cosec}^2\left(\frac{\pi a}{2b}\right) \qquad \ldots.(2)$$

as the final equation for the z/w transformation.

With the particular choice of ζ-plane configuration we can immediately obtain the ζ/w transformation by putting $a = b$ in Equation (2) giving

$$w = 1 + \sinh^2\left(\frac{\pi\zeta}{2b}\right) \qquad \ldots.(3)$$

and now, by eliminating w from these two equations, obtain the final ζ/z transformation as

$$\sinh\left(\frac{\pi z}{2b}\right) = \sin\left(\frac{\pi a}{2b}\right)\sinh\left(\frac{\pi\zeta}{2b}\right) \qquad \ldots.(4)$$

Having obtained the transformation the two parts of the problem can now be solved:

(*a*) In the gap, $z = jy$ and $\zeta = j\eta$. Since the ζ-plane representation is parallel plate the potential in the ζ-plane is just given by

$$\phi = V\left(\frac{\eta}{b}\right)$$

and hence, from Equation (4),

$$\sin\left(\frac{\pi y}{2b}\right) = \sin\left(\frac{\pi a}{2b}\right)\sin\left(\frac{\pi\phi}{2V}\right)$$

giving the potential function in the gap as

$$\phi = \frac{2V}{\pi}\sin^{-1}\left\{\frac{\sin\left(\frac{\pi y}{2b}\right)}{\sin\left(\frac{\pi a}{2b}\right)}\right\}$$

(*b*) Since the capacitance of the z-plane configuration is the same as that of the ζ-plane configuration there is no difficulty in principle in the calculation of capacitance. The only problem in this case is that the capacitance of both is infinite. We are, however, asked to calculate excess capacitance which involves the difference of two infinite quantities.

Let us assume that the z-plane plates are terminated at $x = R$, where R is very large. The equivalent parallel plate capacitance C_p without the septum would then be given by

$$C_p = \frac{\epsilon R}{b} \quad \text{F/m}$$

In order to determine the capacitance with the septum (C_s) we require to find where the ζ-plane parallel plate system terminates, that is, if $z = R + \mathrm{j}y$ what is ζ?

Suppose that we first of all look at the point $z = R + \mathrm{j}0$ with a corresponding $\zeta = \xi + \mathrm{j}0$.

Then, since R is very large

$$\sinh\left(\frac{\pi R}{2b}\right) \approx \tfrac{1}{2} \exp\left(\frac{\pi R}{2b}\right)$$

is also large and hence ξ is large and we can put

$$\sinh\left(\frac{\pi \xi}{2b}\right) \approx \tfrac{1}{2} \exp\left(\frac{\pi \xi}{2b}\right)$$

Thus, from the z/ζ equation we have approximately

$$\frac{\pi R}{2b} \approx \frac{\pi \xi}{2b} + \ln\left\{\sin\left(\frac{\pi a}{2b}\right)\right\}$$

or

$$\xi \approx R + \frac{2b}{\pi} \ln\left\{\operatorname{cosec}\left(\frac{\pi a}{2b}\right)\right\}$$

Now

$$C_s = \left(\frac{\epsilon \xi}{b}\right)$$

and therefore the excess capacitance (or strictly one half of it since we are only using one half of the system on the z-plane) C' is given by

$$C' = C_s - C_p = \frac{\epsilon}{b}(\xi - R) = \frac{2\epsilon}{\pi} \ln\left\{ \operatorname{cosec}\left(\frac{\pi a}{2b}\right)\right\}$$

If R is now allowed to tend to infinity this C' represents the final value of one half of the excess capacitance.

The student should examine the termination of the ζ-plane at $\zeta = \xi + \mathrm{j}b$ to convince himself that the above process is valid when $R \to \infty$.

3.4. From a consideration of the symmetry it is apparent that the problem can be reduced to one quadrant and the transformation we seek is illustrated in *Figure 3.14.* The w-plane configuration is mentioned in the Introduction to Chapter 3 and can be regarded as immediately soluble in that the

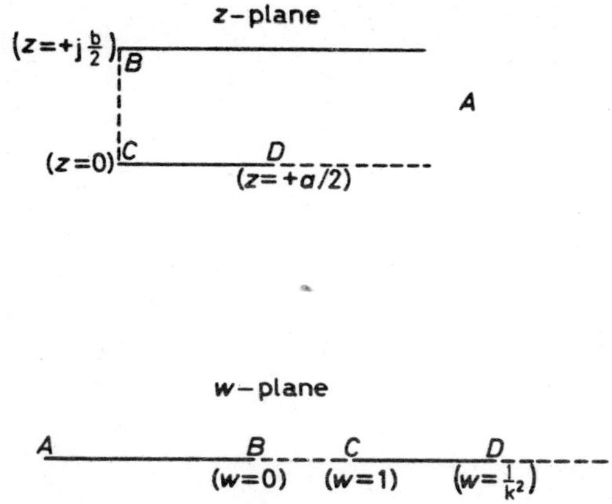

Figure 3.14.

capacitance (appropriate to the upper half-plane) is $\epsilon K'(k)/K(k)$. Since this is one quarter of the total capacitance of the system and is already in the form required our solution requires us to deduce the value of k in terms of the physical dimensions a and b.

The Schwarz-Christoffel transformation is

$$\frac{\mathrm{d}z}{\mathrm{d}w} = \frac{P}{\sqrt{\{w(w-1)\}}} = \frac{P}{\sqrt{\{(w-\tfrac{1}{2})^2 - (\tfrac{1}{2})^2\}}}$$

and if

$$w - \tfrac{1}{2} = \tfrac{1}{2}\cosh\theta$$

then

$$z = P\theta + Q$$

The table of point correspondence is as follows.

z	$j\frac{b}{2}$	0	$\frac{a}{2}$
w	0	1	$1/k^2$
θ	$j\pi$	0	

From the values at z = 0 and $z = jb/2$ it is easily shown that $P = b/2\pi$, $Q = 0$.
The w/z transformation equation is therefore

$$w = \tfrac{1}{2} + \tfrac{1}{2}\cosh\left(\frac{2\pi z}{b}\right) = \cosh^2\left(\frac{\pi z}{b}\right)$$

and from the values at $z = a/2$ we find

$$k = \operatorname{sech}\left(\frac{\pi a}{2b}\right)$$

In order to solve the second part of the problem assume that (a/b) is large enough for k to be regarded as small. Then, using the approximate expression given in the hint,

$$C \approx \frac{8\epsilon}{\pi}\ln\left\{4\cosh\frac{\pi a}{2b}\right\}$$

Now, if (a/b) is large enough to allow us to replace the hyperbolic function by its exponential approximation

$$\cosh\left(\frac{\pi a}{2b}\right) \approx \tfrac{1}{2}\exp\left(\frac{\pi a}{2b}\right)$$

we then obtain

$$C = \frac{8\epsilon}{\pi}\ln\left\{2\exp\left(\frac{\pi a}{2b}\right)\right\}$$

$$= \frac{8\epsilon}{\pi}\ln 2 + \frac{4\epsilon a}{b}$$

The second term in this expression is just the 'parallel plate' value of plates of width a separated by a distance of $b/2$ (there are two of them, of course, due to both sides of the inner strip). The first term can be interpreted as the capacitance of the edge effect, each edge contributing an amount $(4\epsilon/\pi)\ln 2$. This is also sometimes called the fringing capacitance.

For his own benefit the student should investigate the range of values over which the approximations made are valid.

3.5. This problem is more complicated than, but similar to, the previous one. It is of value because it illustrates a new aspect of the use of symmetry. Following the hint given, if the conductors are at the same potential, one can be regarded as the image of the other in a magnetic wall. If they are opposite potentials then they are images in an electric wall. Applying the same criteria to the circuit model of *Figure 3.3*(*b*), if the conductors are at the same potential they each have a capacitance C_p to earth, the coupling capacitance C_c being inoperative. (C_c has no potential difference across it and is storing no charge.) In the case of opposite potentials, C_c can be replaced by two capacitors of value $2C_c$ each, connected in series and the centre connection of these two will be at zero or ground potential. Each line, therefore, has a capacitance of $(C_p + 2C_c)$ to earth.

The overall problem can be regarded as being composed of two separate parts, one with common potentials and the other with opposite potentials. These two parts are called the 'even mode' and 'odd mode' representations respectively. The two reduced problems together with the required w-plane configurations are shown in *Figure 3.15* (for the even mode) and *Figure 3.16* (for the odd mode). Both will be recognised as being similar transformations to that of the previous problem and it is clear that if we can calculate k_e and k_0 the complete solution can be obtained.

(*a*) *Even mode*

The transformation is determined from

$$\frac{dz}{dw} = \frac{P}{\sqrt{\{w(w - w_1)\}}} = \frac{P}{\sqrt{\left\{\left(w - \frac{w_1}{2}\right)^2 - \left(\frac{w_1}{2}\right)^2\right\}}}$$

and letting

$$w - \frac{w_1}{2} = \frac{w_1}{2}\cosh\theta$$

we obtain

$$z = P\theta + Q$$

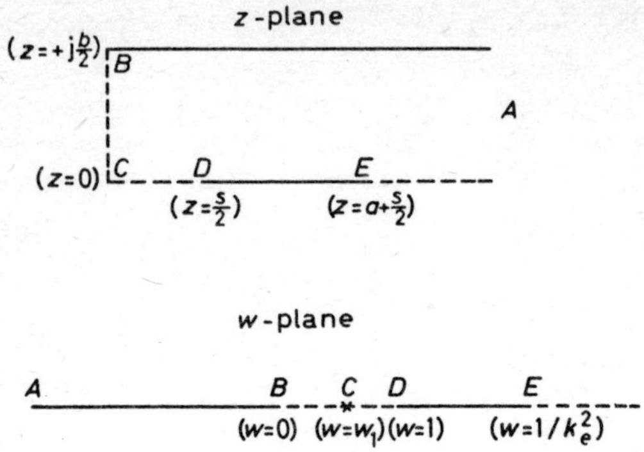

Even mode

Figure 3.15.

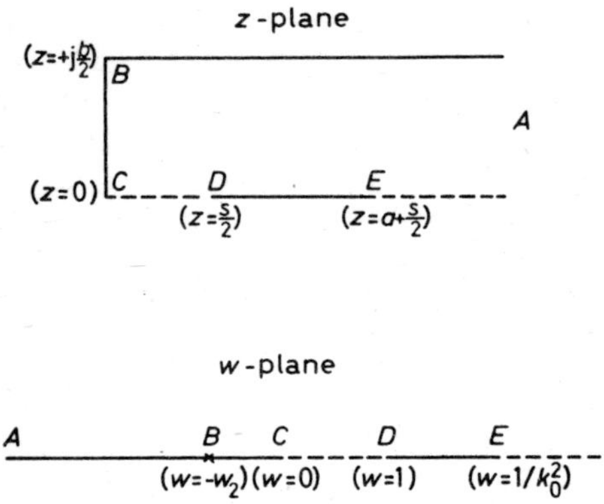

Odd mode

Figure 3.16.

The point correspondence table is as follows.

z	$j\frac{b}{2}$	0	$\frac{s}{2}$	$a+\frac{s}{2}$
w	0	w_1	1	$1/k_e^2$
θ	$j\pi$	0		

and from the values at $z = 0$ and $z = jb/2$ we find that

$$P = b/2\pi \quad \text{and} \quad Q = 0$$

whence

$$w = \frac{w_1}{2}\left\{1 + \cosh\left(\frac{2\pi z}{b}\right)\right\} = w_1 \cosh^2\left(\frac{\pi z}{b}\right)$$

Now since $w = 1$ when $z = s/2$,

$$w_1 = \operatorname{sech}^2\left(\frac{\pi s}{2b}\right)$$

and the complete transformation is

$$w = \operatorname{sech}^2\left(\frac{\pi s}{2b}\right) \cosh^2\left(\frac{\pi z}{b}\right)$$

The values at the fourth point of correspondence give

$$k_e = \cosh\left(\frac{\pi s}{2b}\right) \operatorname{sech}\left[\frac{\pi(s + 2a)}{2b}\right]$$

and this mode is solved.

(b) Odd mode

The procedure is identical, only the values being different. The transformation is determined from

$$\left.\begin{aligned} \frac{dz}{dw} &= \frac{P}{\sqrt{\{w(w + w_2)\}}} \\ w + \frac{w_2}{2} &= \frac{w_2}{2}\cosh\theta \\ z &= P\theta + Q \end{aligned}\right\}$$

with the point correspondence:

z	$j\frac{b}{2}$	0	$\frac{s}{2}$	$a+\frac{s}{2}$
w	$-w_2$	0	1	$1/k_0^2$
θ	$j\pi$	0		

This leads to

$$w = \operatorname{cosech}^2\left(\frac{\pi s}{2b}\right)\sinh^2\left(\frac{\pi z}{b}\right)$$

and, finally,

$$k_0 = \sinh\left(\frac{\pi s}{2b}\right)\operatorname{cosech}\left[\frac{\pi(s+2a)}{2b}\right]$$

Relating the results to the circuit parameters and remembering that we have dealt with one *half* of each line give the equations

$$C_p = 2\epsilon\frac{K'(k_e)}{K(k_e)};\quad C_p + 2C_c = 2\epsilon\frac{K'(k_0)}{K(k_0)}$$

and the final result

$$\left.\begin{aligned} C_p &= 2\epsilon\frac{K'(k_e)}{K(k_e)} \\ C_c &= \epsilon\left\{\frac{K'(k_0)}{K(k_0)} - \frac{K'(k_e)}{K(k_e)}\right\} \end{aligned}\right\}$$

where k_e and k_0 have the values calculated in terms of the physical dimensions.

The far-sighted student who realises that if he lets the dimension s tend to infinity each mode should reduce to Problem 3.4 and therefore the answers should be the same under this limiting process may be disturbed to find an apparent discrepancy. As s becomes very large the limiting values of k_e and k_0 become

$$k_e = k_0 = \exp\left\{-\frac{\pi a}{b}\right\}$$

and C_c tends to zero as it should, C_p being given by

$$C_p = 2\epsilon\,\frac{K'(k)}{K(k)}$$

with

$$k = \exp\left\{-\frac{\pi a}{b}\right\}$$

That this is the same as the answer to Problem **3.4** can be shown via the properties of the elliptic integrals. This mathematical exercise is outside the purpose of this book but the student may care to consider the answer he would have obtained to Problem **3.4** had the analysis been done, not on a quadrant, but on the upper (or lower) *half* of the physical system.

3.6. The geometry of this problem is shown in *Figure 3.17.* Let the inner

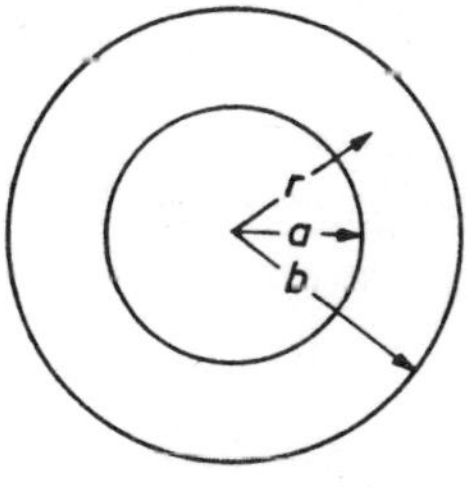

Figure 3.17.

sphere be at potential V_0 and the outer sphere be earthed. The potential ϕ is to vary linearly between $r = a$ and $r = b$ and hence

$$\phi = V_0\left\{\frac{b-r}{b-a}\right\}$$

As mentioned in the Introduction to Chapter 3 a variational expression is obtained for this type of problem by calculation of the stored energy U and application of

$$C = 2U/V_0^2$$

Since

$$U = \int \tfrac{1}{2}\mathbf{E} \,.\, \mathbf{D}\, \mathrm{d}V = \frac{\epsilon}{2} \int |\nabla\phi|^2\, \mathrm{d}V$$

we have to calculate

$$C = \frac{\epsilon}{V_0^2} \int |\nabla\phi|^2\, \mathrm{d}V$$

In the present case ϕ is a function of r alone and therefore

$$\nabla\phi = \mathbf{a}_r \frac{\partial\phi}{\partial r} = -\mathbf{a}_r \frac{V_0}{(b-a)}$$

whence

$$C = \frac{\epsilon}{(b-a)^2} \int_a^b 4\pi r^2\, \mathrm{d}r = \frac{4\pi\epsilon}{(b-a)} \left\{ \frac{b^2 + ab + a^2}{3} \right\}$$

The ratio of this value of capacitance to the correct value C_T is

$$C/C_T = \frac{1}{3} \left\{ \frac{b}{a} + 1 + \frac{a}{b} \right\}$$

and from this we can calculate a table of some representative values as shown below.

b/a	1	1·5	2	4	8
C/C_T	1·00	1·06	1·17	1·75	3·04

The accuracy is not particularly good for $b/a > 1{\cdot}5$ although even at $b/a = 2$ it is only 17% in error. When we consider the crudeness of the approximation to the potential function we have used it is perhaps rather surprising that the accuracy is as good as it is. Let us now consider an improvement. Suppose that we choose a potential function which varies quadratically with r. Let

$$\phi = V_0 \left\{ \alpha \left(\frac{b-r}{b-a} \right) + \beta \left(\frac{b-r}{b-a} \right)^2 \right\}$$

where, in order to obtain $\phi = V_0$ at $r = a$ the constants α and β must satisfy $\alpha + \beta = 1$.

Repeating the previous calculation now produces

$$C = \frac{\epsilon}{(b-a)^2} \int_a^b 4\pi r^2 \left\{ \alpha + 2\beta \left(\frac{b-r}{b-a} \right) \right\}^2 \mathrm{d}r$$

The integration is trivial, although rather lengthy, and when this is carried out, the parameter α put equal to $(1 - \beta)$, the value of capacitance becomes

$$C = \frac{4\pi\epsilon}{3(b-a)} \left\{ (a^2 + ab + b^2) - \beta(b^2 - a^2) + \frac{\beta^2}{5}(2a^2 + ab + 2b^2) \right\}$$

We still have the arbitrary constant β at our disposal. Now comes a rather important point. Since the capacitance calculated by this method is *always greater* than the true value, and this is true whatever the value of β, the obvious choice of β is that value which minimizes the above capacitance. By differentiation, therefore, we obtain

$$\beta = \frac{5(b^2 - a^2)}{2(2a^2 + ab + 2b^2)}$$

Note that, in effect, we are choosing a different value of β for each different b/a. Inserting this value and defining a quantity R by

$$R = \frac{b}{a} + \frac{a}{b}$$

it is straightforward algebra to show that

$$\frac{C}{C_T} = \frac{(R^2 + 4R + 8)}{4(2R + 1)}$$

We can now create a table as before.

b/a	1	1·5	2	4	8
C/C_T	1·00	1·00	1·01	1·13	1·54

The improvement in accuracy is apparent.

3.7. It is sufficient to consider one quarter of the line as shown in *Figure 3.18.* The linear potential variation is assumed along the line AA' and this

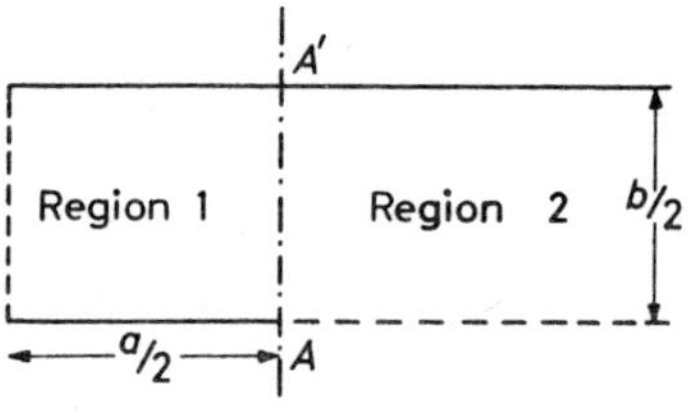

Figure 3.18.

splits the whole region into two parts. In Region 1 the linear assumption enforces an ideal parallel plate field in the region. In Region 2 the potential must be obtained by the method of separation of variables.

Region 1

Since we are enforcing an ideal parallel plate field in this region the capacitance, calculated from the energy or any other means will be

$$C_1 = \frac{\epsilon a}{b}$$

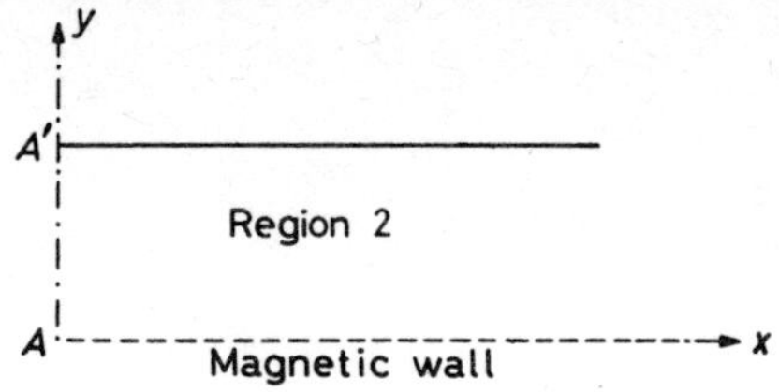

Figure 3.19.

Region 2

First of all attach a coordinate system to the region as shown in *Figure 3.19.* The line $y = 0$ is a magnetic wall. On AA' we assume a linear potential function of the form

$$\phi(y) = V_0\left\{1 - \frac{2y}{b}\right\}$$

It is easily shown by the separation of variables that the potential function in the strip which satisfies the boundary conditions:

(*i*) $\phi = 0$ at $x = \infty$
(*ii*) $\partial\phi/\partial y = 0$ at $y = 0$
(*iii*) $\phi = 0$ at $y = b/2$

is

$$\phi(x, y) = \sum_{n=0}^{\infty} A_n \cos\left\{\frac{(2n+1)\pi y}{b}\right\} \exp\left\{-\frac{(2n+1)\pi x}{b}\right\}$$

It remains to satisfy the condition at $x = 0$, that is

$$\phi(0, y) = V_0\left\{1 - \frac{2y}{b}\right\} = \sum_{n=0}^{\infty} A_n \cos\left\{\frac{(2n+1)\pi y}{b}\right\}$$

and from the usual methods of Fourier analysis we find

$$A_n = \frac{8V_0}{\pi^2} \cdot \frac{1}{(2n+1)^2}$$

We have now completely determined the potential function and are in a position to calculate the energy.

Now

$$\frac{\partial \phi}{\partial x} = -\frac{\pi}{b}\sum_{n=0}^{\infty}(2n+1)A_n\cos\left\{\frac{(2n+1)\pi y}{b}\right\}\exp\left\{-\frac{(2n+1)\pi x}{b}\right\}$$

and

$$\frac{\partial \phi}{\partial y} = -\frac{\pi}{b}\sum_{n=0}^{\infty}(2n+1)A_n\sin\left\{\frac{(2n+1)\pi y}{b}\right\}\exp\left\{-\frac{(2n+1)\pi x}{b}\right\}$$

and therefore

$$|\nabla\phi|^2 = \left(\frac{\pi}{b}\right)^2\sum_{m=0}^{\infty}\sum_{n=0}^{\infty}(2m+1)(2n+1)A_mA_n\cos\left\{\frac{2(m-n)\pi y}{b}\right\}$$

$$\times\exp\left\{-\frac{2(m+n+1)\pi x}{b}\right\}$$

after various terms in the double series have been combined.

We must now form

$$U = \frac{\epsilon}{2}\int_{x=0}^{\infty}\int_{y=0}^{b/2}|\nabla\phi|^2\,dx\,dy$$

and if the integration with respect to y is carried out first, it will be found that most of the terms vanish, leaving

$$\int_{y=0}^{b/2}|\nabla\phi|^2\,dy = \frac{\pi^2}{2b}\sum_{n=0}^{\infty}(2n+1)^2A_n^2\exp\left\{-\frac{2(2n+1)\pi x}{b}\right\}$$

The integration with respect to x can now be carried out giving

$$U = \frac{\pi\epsilon}{8}\sum_{n=0}^{\infty}(2n+1)A_n^2 = \frac{8}{\pi^3}\epsilon V_0^2\sum_{n=0}^{\infty}\frac{1}{(2n+1)^3}$$

whence the capacitance for Region 2 is

$$C_2 = \frac{2U}{V_0^2} = \frac{16\epsilon}{\pi^3} \sum_{n=0}^{\infty} \frac{1}{(2n+1)^3}$$

The summation is a standard form with a value of 1·0518 (it can in any case be easily calculated since the series is rapidly convergent) and we obtain finally

$$C_2 = 0{\cdot}543\epsilon$$

The capacitance for the system of *Figure 3.18* is the sum of C_1 and C_2 and this is one quarter of the capacitance of the whole line. Thus an upper bound to the capacitance of the strip transmission line is

$$C = 4(C_1 + C_2) = \frac{4\epsilon a}{b} + 2{\cdot}172\epsilon$$

This can be compared with the exact answer found by conformal transformation in Problem 3.4 but a more straightforward comparison can be made with the approximate answer obtained in that problem. Both answers contain the same parallel plate term, the only difference being in the value of edge capacitance. In the present problem this is 2·172ϵ compared with the 1·764ϵ obtained before. Obviously for large values of a/b the differences between the two *total* capacitances is negligible although the edge capacitance difference appears large.

The student should try to improve the approximation by the use of a quadratic law. This will require an analysis of Region 1 by separation of variables in addition to Region 2.

3.8. Using symmetry we choose to consider one quarter of the line as shown in *Figure 3.20*, assuming a linear variation of potential along AA'.

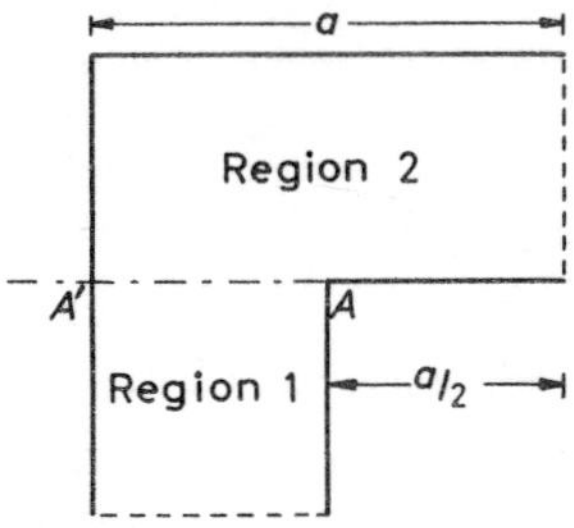

Figure 3.20.

This splits the quarter line into two regions as designated, Region 1 being of parallel plate form and having a capacitance $C_1 = \epsilon$ F/m.

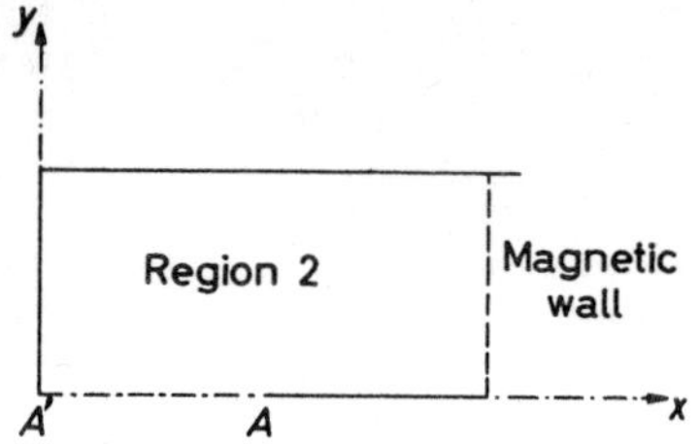

Figure 3.21.

Consider Region 2 with the coordinate system as shown in *Figure 3.21.* The boundary conditions are:

(*i*) $y = a/2, \phi = 0$
(*ii*) $x = 0, \phi = 0$
(*iii*) $x = a, \partial\phi/\partial x = 0$ (magnetic wall)
(*iv*) $y = 0, \phi = V_0\left(\frac{2x}{a}\right)$ for $0 \leqslant x \leqslant (a/2)$
and $\phi = V_0$ for $(a/2) \leqslant x \leqslant a$

Using the normal technique for the separation of variables, the first three of these conditions are satisfied by a potential function

$$\phi(x, y) = \sum_{n=0}^{\infty} A_n \sin\left\{\frac{(2n+1)\pi x}{2a}\right\} \sinh\left\{\frac{(2n+1)\pi\left(\frac{a}{2} - y\right)}{2a}\right\}$$

where A_n can be calculated from

$$\phi(x, 0) = \sum_{n=0}^{\infty} A_n \sinh\left\{\frac{(2n+1)\pi}{4}\right\} \sin\left\{\frac{(2n+1)\pi x}{2a}\right\}$$

whence

$$A_n \sinh\left\{\frac{(2n+1)\pi}{4}\right\} = \frac{16V_0}{(2n+1)^2\pi^2} \sin\left\{\frac{(2n+1)\pi}{4}\right\}$$

carrying out the Fourier inversion.

In order to use the variational process we must now calculate the energy in the region. Now

$$\frac{\partial\phi}{\partial x} = \left(\frac{\pi}{2a}\right) \sum_{n=0}^{\infty} (2n+1) A_n \cos\left\{\frac{(2n+1)\pi x}{2a}\right\} \sinh\left\{\frac{(2n+1)\pi\left(\frac{a}{2}-y\right)}{2a}\right\}$$

$$\frac{\partial\phi}{\partial y} = -\left(\frac{\pi}{2a}\right) \sum_{n=0}^{\infty} (2n+1) A_n \sin\left\{\frac{(2n+1)\pi x}{2a}\right\} \cosh\left\{\frac{(2n+1)\pi\left(\frac{a}{2}-y\right)}{2a}\right\}$$

and it is relatively easy although rather lengthy to show that

$$U = \frac{\pi\epsilon}{16} \sum_{n=0}^{\infty} (2n+1) A_n^2 \sinh\left\{\frac{(2n+1)\pi}{2}\right\}$$

We can now substitute for A_n and, using the two relations,

$$\sin^2\left\{\frac{(2n+1)\pi}{4}\right\} = \tfrac{1}{2}$$

$$\sinh\left\{\frac{(2n+1)\pi}{2}\right\} = 2 \sinh\left\{\frac{(2n+1)\pi}{4}\right\} \cosh\left\{\frac{(2n+1)\pi}{4}\right\}$$

derive

$$U = \frac{16\epsilon V_0^2}{\pi^3} \sum_{n=0}^{\infty} \frac{\coth\left\{\frac{(2n+1)\pi}{4}\right\}}{(2n+1)^3}$$

whence the value of C_2 is given by

$$C_2 = \frac{32\epsilon}{\pi^3} \sum_{n=0}^{\infty} \frac{\coth\left\{\frac{(2n+1)\pi}{4}\right\}}{(2n+1)^3}$$

The series is again rapidly convergent with a sum of 1·578 leading to

$$C_2 = 1{\cdot}625\epsilon$$

The capacitance of one quarter of the line, $(C_1 + C_2)$, is therefore $2{\cdot}625\epsilon$ and the total capacitance C is four times this, whence,

$$C = 10{\cdot}500\epsilon$$

Evaluating this for free space

$$C = 92{\cdot}925 \times 10^{-12} \qquad \text{F/m}$$

a value which compares very favourably with the exact value of 90·4 pF/m.

3.9. This important property is proved by use of the variational principle. Suppose that we know the potential for the problem with conductors at S_2 and S_3. For the problem involving conductors at S_1 and S_3 let us assume this known potential for the space between the conductor at S_3 and the now imaginary surface S_2. In the space between S_2 and S_1 we assume that the potential is constant.

In the calculation of energy for the second problem the region between S_3 and S_2 will store exactly the same energy as in the problem involving a conductor at S_2. The region between S_2 and S_1 is (under the assumption) at a constant potential and is therefore field free, storing no energy. Thus the capacitance for S_3 and S_1 calculated by the variational method will be just C_{23} and this value is therefore an upper bound to C_{13}.

Therefore

$$C_{23} \geqslant C_{13}$$

Now, as long as the surfaces S_2 and S_1 do not coincide, we can choose another surface S_2' between S_2 and S_1 and, by a similar argument derive

$$C_{23} \geqslant C_{23}' \geqslant C_{13}$$

Thus, unless the capacitance is the same for *any* surface inside S_2, then

$$C_{23} > C_{13}$$

3.10. This is a good example of a problem in which a combination of methods can be used to obtain both an upper and a lower bound to capacitance. Obviously if both bounds can be obtained we know the maximum possible error in any estimate of the capacitance.

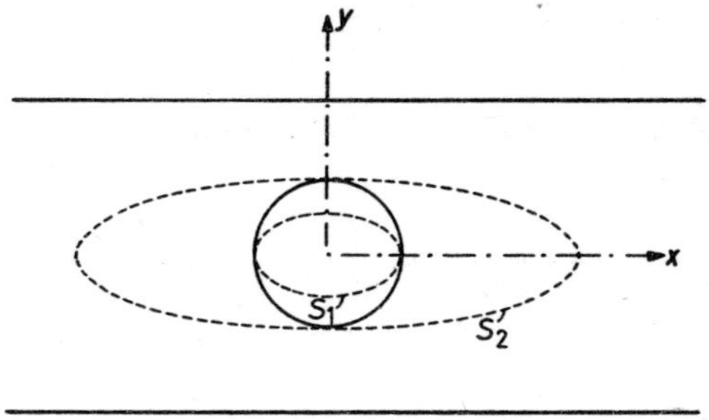

Figure 3.22.

The calculation is illustrated with reference to *Figure 3.22* in which the ground planes and inner conductor are shown together with two of the equipotential surfaces of Problem 3.1 (designated as S_1 and S_2). By the result of Problem 3.9 the capacitance C_2 of a conductor at S_2 (to ground) is greater than the capacitance C of the given transmission line. Similarly, the capacitance C_1 of a conductor at S_1 (to ground) is less than the capacitance C. Thus C_1 and C_2 are lower and upper bounds, respectively, to the capacitance C. It now remains to calculate C_1 and C_2.

From Problem 3.1 the potential function of a line charge between the ground planes is

$$\phi(x, y) = \frac{\sigma}{4\pi\epsilon} \ln \left\{ \frac{\cosh\left(\frac{\pi x}{b}\right) + \cos\left(\frac{\pi y}{b}\right)}{\cosh\left(\frac{\pi x}{b}\right) - \cos\left(\frac{\pi y}{b}\right)} \right\}$$

Thus the potential at the point $(a, 0)$, i.e. for the surface S_1, is

$$\phi_1 = \frac{\sigma}{4\pi\epsilon} \ln \left\{ \frac{\cosh\left(\frac{\pi a}{b}\right) + 1}{\cosh\left(\frac{\pi a}{b}\right) - 1} \right\} = \frac{\sigma}{2\pi\epsilon} \ln \left\{ \coth\left(\frac{\pi a}{2b}\right) \right\}$$

and the potential at the point $(0, a)$, i.e. for the surface S_2, is

$$\phi_2 = \frac{\sigma}{4\pi\epsilon} \ln \left\{ \frac{1 + \cos\left(\frac{\pi a}{b}\right)}{1 - \cos\left(\frac{\pi a}{b}\right)} \right\} = \frac{\sigma}{2\pi\epsilon} \ln \left\{ \cot\left(\frac{\pi a}{2b}\right) \right\}$$

Since $C_1 = \sigma/\phi_1$ and $C_2 = \sigma/\phi_2$ we obtain the bounds

$$\frac{1}{\ln\left\{\coth\left(\frac{\pi a}{2b}\right)\right\}} < \frac{C}{2\pi\epsilon} < \frac{1}{\ln\left\{\cot\left(\frac{\pi a}{2b}\right)\right\}}$$

The following table gives some sample values

$2a/b$	0·2	0·4	0·6	0·8
$C_1/2\pi\epsilon$	0·538	0·840	1·216	1·707
$C_2/2\pi\epsilon$	0·543	0·890	1·483	3·133
Mean	0·540	0·865	1·350	2·420
Error (%)	0·5	2·9	9·8	29·0

CHAPTER 4

4.1. Maxwell's equations are:

$$\nabla \times \mathbf{E} = -\frac{\partial \mathbf{B}}{\partial t}$$

$$\nabla . \mathbf{D} = \rho$$

$$\nabla \times \mathbf{H} = \mathbf{J} + \frac{\partial \mathbf{D}}{\partial t}$$

$$\nabla . \mathbf{B} = 0$$

Assuming free space conditions, i.e. $\rho = \mathbf{J} = 0$; $\mathbf{B} = \mu_0\mathbf{H}$, $\mathbf{D} = \epsilon_0\mathbf{E}$, these become

$$\nabla \times \mathbf{E} = -\mu_0\frac{\partial \mathbf{H}}{\partial t} \qquad \ldots .(1)$$

$$\nabla . \mathbf{E} = 0 \qquad \ldots .(2)$$

$$\nabla \times \mathbf{H} = \epsilon_0\frac{\partial \mathbf{E}}{\partial t} \qquad \ldots .(3)$$

$$\nabla . \mathbf{H} = 0 \qquad \ldots .(4)$$

Now taking the curl of the curl of **H** and using Equations (1) and (4) together with the vector identity given, we obtain

$$\nabla \times \nabla \times \mathbf{H} = \epsilon_0\frac{\partial}{\partial t}(\nabla \times \mathbf{E}) = -\mu_0\epsilon_0\frac{\partial^2 \mathbf{H}}{\partial t^2} = \nabla(\nabla . \mathbf{H}) - \nabla^2\mathbf{H}$$

$$\therefore \qquad \nabla^2\mathbf{H} = \mu_0\epsilon_0\frac{\partial^2 \mathbf{H}}{\partial t^2}$$

Similarly,

$$\nabla \times \nabla \times \mathbf{E} = -\mu_0\frac{\partial}{\partial t}(\nabla \times \mathbf{H}) = -\mu_0\epsilon_0\frac{\partial^2 \mathbf{E}}{\partial t^2} = \nabla(\nabla . \mathbf{E}) - \nabla^2\mathbf{E}$$

$$\therefore \qquad \nabla^2\mathbf{E} = \mu_0\epsilon_0\frac{\partial^2 \mathbf{E}}{\partial t^2}$$

4.2.

$$\nabla^2 \mathbf{H} = \mu\sigma \frac{\partial \mathbf{H}}{\partial t} + \mu\epsilon \frac{\partial^2 \mathbf{H}}{\partial t^2} \qquad \ldots .(1)$$

Assuming a time dependence of the form $e^{j\omega t}$, then we can replace $\partial/\partial t$ by $j\omega$ and $\partial^2/\partial t^2$ by $-\omega^2$. Hence

$$\nabla^2 \mathbf{H} = (j\omega\mu\sigma - \omega^2\mu\epsilon)\mathbf{H} = \gamma^2 \mathbf{H} \qquad \ldots .(2)$$

For convenience let us assume that the plane wave is propagating in the x direction and that **H** lies along the y axis. Then

$$\nabla^2 H_y = \frac{\partial^2 H_y}{\partial x^2} = \gamma^2 H_y \qquad \ldots .(3)$$

Here we have used the fact that the wave is plane, i.e. $(\partial/\partial y) = (\partial/\partial z) = 0$ and uniform. The solution of Equation (3) is

$$H_y = A e^{-\gamma x} + B e^{\gamma x}$$

where γ is called the propagation constant, i.e.

$$\gamma^2 = \omega\mu[j\sigma - \omega\epsilon] = \omega\mu\sigma[j - \omega\epsilon/\sigma]$$

Now by a 'good conductor' we mean one for which the conduction current density greatly exceeds the displacement current density, i.e. $(\omega\epsilon/\sigma) \ll 1$. In this case then, $\gamma^2 \doteqdot j\omega\mu\sigma$

and $$\gamma = \frac{1 + j}{\sqrt{2}} \sqrt{(\omega\mu\sigma)} = (1 + j)\sqrt{(\pi f \mu\sigma)} \qquad \ldots .(4)$$

For the constants given in our particular problem we have

$$\frac{\omega\epsilon}{\sigma} = \frac{(2\pi 10^6)(8{\cdot}854 \times 10^{-12} . 80)}{4} \approx 0{\cdot}001$$

Thus at this frequency, sea water can be considered a good conductor. Now

$$\sqrt{(\pi f \mu \sigma)} = \sqrt{(\pi 10^6 4\pi 10^{-7} 4)} = \frac{4\pi}{\sqrt{10}} = 3{\cdot}98$$

Let d be the depth at which the magnetic field is 0·01 of its surface value, then

$$e^{-3{\cdot}98d} = 0{\cdot}01$$

$$\therefore \qquad d = 1{\cdot}15 \text{ m}$$

4.3. Let us assume that the dielectric is linear, lossless and isotropic (for the simple reason that any other assumptions would take the problem outside the scope of this book) and that the fields have a time dependence of the form $e^{j\omega t}$. This assumed time dependence is no real restriction since any arbitrary time dependence can be reduced to the sum of such terms by means of a Fourier analysis; besides it keeps the analysis simple.

Under these assumptions then, Maxwell's equations are:

$$\nabla \times \mathbf{E} = -j\omega\mu\mathbf{H} \qquad \ldots . (1a)$$

$$\nabla \times \mathbf{H} = j\omega\epsilon\mathbf{E} \qquad \ldots . (1b)$$

$$\nabla . \mathbf{H} = 0 \qquad \ldots . (1c)$$

$$\nabla . \mathbf{E} = 0 \qquad \ldots . (1d)$$

Now for plane waves, the only space variations are with respect to l (see *Figure 4.4*)

$$\therefore \qquad \nabla = \mathbf{n}\frac{\partial}{\partial l} \qquad \ldots . (2)$$

hence, putting $\mathbf{n}(\partial/\partial l)$ for ∇ into the divergence relations, we see that both **E** and **H** are perpendicular to **n**. Thus the curl equations simplify to:

$$\mathbf{n} \times \frac{\partial \mathbf{E}}{\partial l} = -j\omega\mu\mathbf{H} \qquad \ldots . (3)$$

$$\mathbf{n} \times \frac{\partial \mathbf{H}}{\partial l} = j\omega\epsilon\mathbf{E} \qquad \ldots . (4)$$

Now

$$\mathbf{n} \times \mathbf{n} \times \frac{\partial \mathbf{E}}{\partial l} = -\frac{\partial \mathbf{E}}{\partial l} = -\mathrm{j}\omega\mu\,(\mathbf{n} \times \mathbf{H})$$

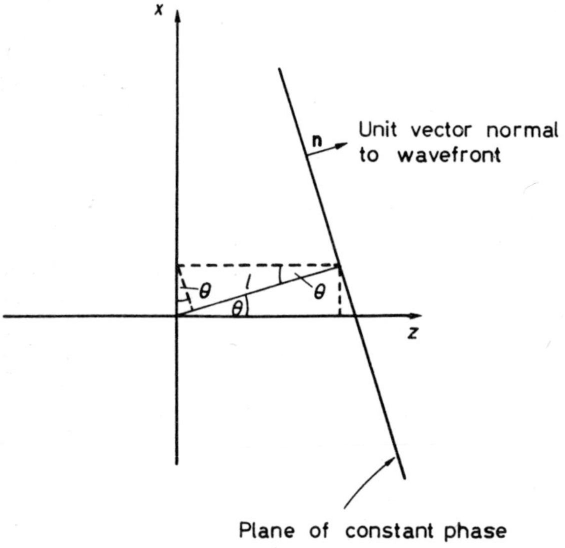

Figure 4.4.

Differentiating both sides with respect to l and substituting from Equation (4) above gives

$$\frac{\partial^2 \mathbf{E}}{\partial l^2} = -\omega^2\mu\epsilon\mathbf{E} = \mathrm{j}^2\beta^2\mathbf{E}$$

Hence

$$\mathbf{E} = \mathscr{E}^{+}\mathrm{e}^{\mathrm{j}(\omega t - \beta l)} + \mathscr{E}^{-}\mathrm{e}^{\mathrm{j}(\omega t + \beta l)} \qquad \ldots .(5)$$

The term associated with $\mathscr{E}^{+}$ represents a wave travelling along the $+l$ axis and that associated with $\mathscr{E}^{-}$ a wave travelling along the $-l$ axis. These are referred to as incident or 'forward' wave and reflected or 'backward' wave respectively.

Using Equation (5) in Equation (3) we get

$$\mathbf{H} = \frac{1}{\sqrt{\mu/\epsilon}}\left[\, \mathbf{n} \times \mathscr{E}^{+} e^{j(\omega t - \beta l)} - \mathbf{n} \times \mathscr{E}^{-} e^{j(\omega t + \beta l)} \right]$$

$$= \mathscr{H}^{+} + \mathscr{H}^{-} \qquad \ldots . (6)$$

from which we see that

$$\frac{\mathbf{n} \times \mathscr{E}^{+}}{\mathscr{H}^{+}} = \frac{\mathbf{n} \times \mathscr{E}^{-}}{\mathscr{H}^{-}} = \sqrt{\frac{\mu}{\epsilon}} = \eta$$

the 'intrinsic impedance' of the medium. We also note that both $\mathscr{E}^{+}$ and $\mathscr{H}^{+}$ are orthogonal to each other and to the direction of propagation. This holds also for $\mathscr{E}^{-}$ and $\mathscr{H}^{-}$. *Figure 4.5* shows the relationship.

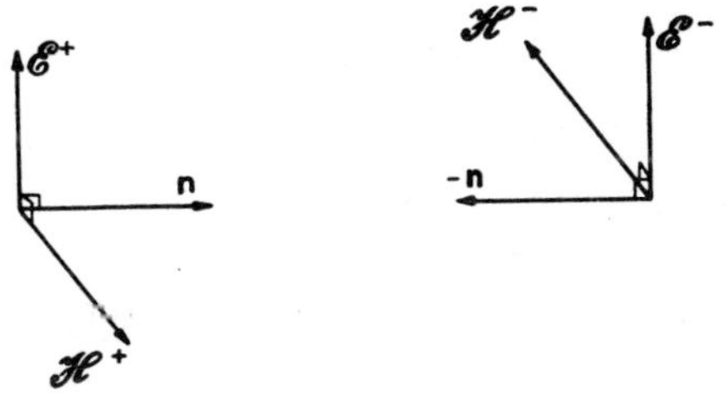

Figure 4.5.

Consider the discontinuity shown in *Figure 4.6* where the incident wave is polarized in the plane of incidence. The boundary conditions that must be satisfied across *s* are that the tangential electric and magnetic fields must be continuous. When this is so, the normal components are automatically satisfied. Now $l = x \sin\theta + z \cos\theta$ and so we see that on the boundary at $z = 0$, the waves in Medium 1 have an exponential factor of the form $e^{j(\omega t - \beta_1 x \sin\theta_1)}$ for the forward wave and $e^{j(\omega t - \beta_1 x \sin\theta_2)}$ for the backward wave and that the waves in Medium 2 have an exponential factor of the form $e^{j(\omega t - \beta_2 x \sin\phi_1)}$ for the forward wave and $e^{j(\omega t - \beta_2 x \sin\phi_2)}$ for the backward wave.

In order that the tangential fields can be matched across *s* for all *t* and *x* it follows that all of these exponential factors must be equal. Hence

$$\theta_1 = \theta_2 = \theta$$

and

$$\phi_1 = \phi_2 = \phi$$

Also, $\beta_1 \sin\theta = \beta_2 \sin\phi$ is the well known Snell's law.

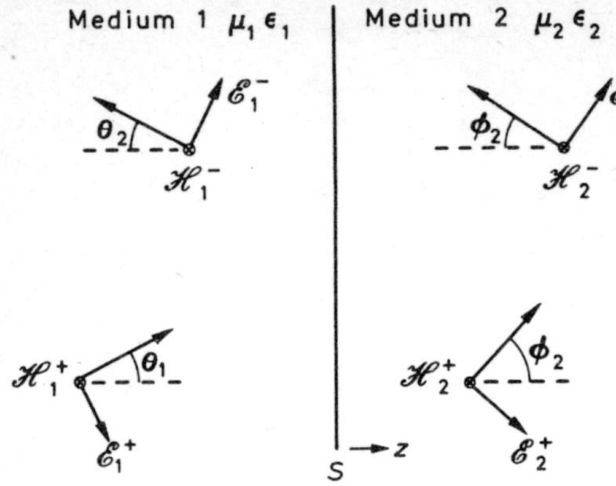

Figure 4.6.

Applying the boundary conditions we have:

$$\mathscr{H}_{s1}^+ + \mathscr{H}_{s1}^- = \mathscr{H}_{s2}^+ + \mathscr{H}_{s2}^- \qquad \ldots.(7)$$

$$(\mathscr{E}_{s1}^+ - \mathscr{E}_{s1}^-)\cos\theta = (\mathscr{E}_{s2}^+ - \mathscr{E}_{s2}^-)\cos\phi \qquad \ldots.(8)$$

Using $\mathscr{E}^+/\mathscr{H}^+ = \mathscr{E}^-/\mathscr{H}^- = \eta$ in Equations (7) and (8) we obtain

$$\begin{bmatrix} \mathscr{H}_{s1}^+ \\ \mathscr{H}_{s1}^- \end{bmatrix} = \begin{bmatrix} \dfrac{\eta_1\cos\theta + \eta_2\cos\phi}{2\eta_1\cos\theta} & \dfrac{\eta_1\cos\theta - \eta_2\cos\phi}{2\eta_1\cos\theta} \\ \dfrac{\eta_1\cos\theta - \eta_2\cos\phi}{2\eta_1\cos\theta} & \dfrac{\eta_1\cos\theta + \eta_2\cos\phi}{2\eta_1\cos\theta} \end{bmatrix} \begin{bmatrix} \mathscr{H}_{s2}^+ \\ \mathscr{H}_{s2}^- \end{bmatrix}$$

$$= \begin{bmatrix} A_{11} & A_{12} \\ A_{21} & A_{22} \end{bmatrix} \begin{bmatrix} \mathscr{H}_{s2}^+ \\ \mathscr{H}_{s2}^- \end{bmatrix}$$

In our particular problem there is no mechanism for setting up the backward wave in the dielectric, thus $\mathscr{H}_{s2}^- = 0$.

Now
$$\mathscr{H}_{s1}^+ = A_{11}\,\mathscr{H}_{s2}^+$$
$$\mathscr{H}_{s1}^- = A_{21}\,\mathscr{H}_{s2}^+$$

If $\mathscr{H}_{s1}^{-}$ is zero, then by conservation of energy $\mathscr{H}_{s2}^{+} \neq 0$, hence $A_{21} = 0$

$$\therefore \qquad \eta_1 \cos\theta = \eta_2 \cos\phi \qquad \ldots .(9)$$

Using Snell's law,

$$\sin^2\phi = \frac{\beta_1{}^2}{\beta_2{}^2} \sin^2\theta = \frac{\omega^2\mu_0\epsilon_0}{\omega^2\mu_0\epsilon_0\epsilon_r} \sin^2\theta = \frac{\sin^2\theta}{\epsilon_r}$$

and from Equation (9) we have

$$\cos^2\phi = \frac{\eta_1{}^2}{\eta_2{}^2} \cos^2\theta = \frac{\mu_0/\epsilon_0}{\mu_0/\epsilon_0\epsilon_r} \cdot \cos^2\theta = \epsilon_r \cos^2\theta$$

$$\therefore \qquad \cos^2\phi + \sin^2\phi = 1 = \frac{\sin^2\theta}{\epsilon_r} + \epsilon_r \cos^2\theta$$

$$\therefore \qquad \epsilon_r \sec^2\theta = \tan^2\theta + \epsilon_r{}^2$$

$$\therefore \qquad \tan^2\theta\,[1 - \epsilon_r] = \epsilon_r[1 - \epsilon_r]$$

$$\therefore \qquad \underline{\theta = \tan^{-1}\sqrt{\epsilon_r}}$$

But $\theta = 56{\cdot}6°$

$$\therefore \qquad \epsilon_r = 2{\cdot}3$$

If the electric vector of the incident wave is now rotated through a small angle, we can consider this new case as the superposition of two simpler waves; one in the plane of incidence and the other perpendicular to it. The first and largest is the same as the case just solved, and this component will pass entirely into the dielectric with no reflection. The component with the electric vector perpendicular to the plane of incidence will suffer some reflection since there is no angle for which this is zero.

To sum up then, the reflected wave will be polarized with its electric vector perpendicular to the plane of incidence and will be small. The transmitted wave will have both components but will largely be polarized in the plane of incidence.

4.4.

$$v_g \equiv \frac{d\omega}{d\beta}$$

$$v_p \equiv \frac{\omega}{\beta}$$

Since $$\omega = \beta v_p$$

Then $$\frac{d\omega}{d\beta} \equiv v_g = v_p + \beta \frac{dv_p}{d\beta}$$

Now $$\beta = \frac{2\pi}{\lambda}$$

$$\therefore \quad \beta \frac{dv_p}{d\beta} = \frac{2\pi}{\lambda} \cdot \frac{dv_p}{d\lambda}\left(-\frac{\lambda^2}{2\pi}\right) = -\lambda \frac{dv_p}{d\lambda}$$

Hence

$$\underline{v_g = v_p - \lambda \frac{dv_p}{d\lambda}} \qquad \ldots .(1)$$

We can also write

$$\beta \frac{dv_p}{d\beta} = \beta \frac{dv_p}{d\omega} v_g = \frac{\omega}{v_p} \frac{dv_p}{d\omega} v_g$$

Hence

$$v_g \left[1 - \frac{\omega}{v_p} \frac{dv_p}{d\omega}\right] = v_p$$

$$\therefore \quad \underline{\frac{1}{v_p} - \frac{1}{v_g} = \frac{\omega}{v_p{}^2} \frac{dv_p}{d\omega}} \qquad \ldots .(2)$$

If α and β are the real and imaginary parts of the propagation factor, then

$$(\alpha + j\beta)^2 = -\omega^2 \mu_0 \epsilon_0 (k_e' - jk_e'')$$

Equating the real and imaginary parts of each side gives:

$$\alpha^2 - \beta^2 = -\omega^2\mu_0\epsilon_0 k_e'$$

and

$$2\alpha\beta = \omega^2\mu_0\epsilon_0 k_e''$$

These are easily solved for α and β and yield:

$$\alpha = \frac{\omega}{c}\sqrt{\frac{k_e'}{2}\left\{\sqrt{1+\left(\frac{k_e''}{k_e'}\right)^2} - 1\right\}}$$

$$\beta = \frac{\omega}{c}\sqrt{\frac{k_e'}{2}\left\{\sqrt{1+\left(\frac{k_e''}{k_e'}\right)^2} + 1\right\}}$$

Since $\quad k_e'' \ll k_e', \alpha \approx 0, \beta \approx \frac{\omega}{c}\sqrt{k_e'}$

Hence $\quad \dfrac{1}{v_p} = \dfrac{\beta}{\omega} \approx \dfrac{\sqrt{k_e'}}{c}$

Now $\quad \underline{v_p \approx c(k_e')^{-1/2}}$

$$\therefore \quad \frac{\mathrm{d}v_p}{\mathrm{d}\omega} \approx -\frac{c}{2}(k_e')^{-3/2}\frac{\mathrm{d}k_e'}{\mathrm{d}\omega} = -\frac{1}{2}\frac{v_p}{k_e'}\frac{\mathrm{d}k_e'}{\mathrm{d}\omega}$$

Therefore from Equation (2)

$$\underline{\frac{1}{v_g} \approx \frac{1}{v_p}\left(1 + \frac{\omega}{2k_e'}\frac{\mathrm{d}k_e'}{\mathrm{d}\omega}\right)}$$

4.5. (*a*) For the monochromatic case, the forward and backward waves have exponential factors of the form

$$\mathrm{e}^{j\omega t - \gamma z} \quad \text{and} \quad \mathrm{e}^{j\omega t + \gamma z}$$

respectively and hence we can see by inspection that

$$\left.\begin{array}{ll} \mathscr{E}^+_{r2} = \mathscr{E}^+_{s2}\,\mathrm{e}^{-\gamma l}, & \mathscr{H}^+_{r2} = \mathscr{H}^+_{s2}\,\mathrm{e}^{-\gamma l} \\ \mathscr{E}^-_{s2} = \mathscr{E}^-_{r2}\,\mathrm{e}^{-\gamma l}, & \mathscr{H}^-_{s2} = \mathscr{H}^-_{r2}\,\mathrm{e}^{-\gamma l} \end{array}\right\} \quad \ldots.(1)$$

$\mathscr{E}^+_{r2}$ refers to the value of $\mathscr{E}^+_2$ at the plane *r*, and similarly for the other components. As is the usual practice, we have suppressed the time dependence since every field component has this same time factor and it gets rather tedious writing it out all the time. It is implied however and should not be forgotten.

Equations (1) can be written in the matrix form:

$$\begin{bmatrix} \mathscr{E}^+_{s2} \\ \mathscr{E}^-_{s2} \end{bmatrix} = \begin{bmatrix} e^{\gamma l} & 0 \\ 0 & e^{-\gamma l} \end{bmatrix} \begin{bmatrix} \mathscr{E}^+_{r2} \\ \mathscr{E}^-_{r2} \end{bmatrix} \qquad \ldots . (2)$$

and an identical expression for the magnetic components.

(*b*) To obtain the wave amplitude matrix of the junction at *s* we need only note that the tangential components of the field across the junction must be continuous, i.e.

$$\left.\begin{aligned} \mathscr{E}^+_{s1} + \mathscr{E}^-_{s1} &= \mathscr{E}^+_{s2} + \mathscr{E}^-_{s2} \\ \text{and} \quad \mathscr{H}^+_{s1} - \mathscr{H}^-_{s1} &= \mathscr{H}^+_{s2} - \mathscr{H}^-_{s2} \end{aligned}\right\} \qquad \ldots . (3)$$

using

$$\frac{\mathscr{E}^+_1}{\mathscr{H}^+_1} = \frac{\mathscr{E}^-_1}{\mathscr{H}^-_1} = \eta_0$$

and

$$\frac{\mathscr{E}^+_2}{\mathscr{H}^+_2} = \frac{\mathscr{E}^-_2}{\mathscr{H}^-_2} = \eta$$

we obtain from Equation (3)

$$\left.\begin{aligned} \mathscr{E}^+_{s1} &= \left(\frac{\eta + \eta_0}{2\eta}\right) \mathscr{E}^+_{s2} + \left(\frac{\eta - \eta_0}{2\eta}\right) \mathscr{E}^-_{s2} \\ \mathscr{E}^-_{s1} &= \left(\frac{\eta - \eta_0}{2\eta}\right) \mathscr{E}^+_{s2} + \left(\frac{\eta + \eta_0}{2\eta}\right) \mathscr{E}^-_{s2} \end{aligned}\right\} \qquad \ldots . (4)$$

which clearly satisfies the matrix equation given.

4.6. Because the dielectric is lossless γ must be imaginary, $j\beta$ say. Also we note that since the slab is in an infinite media, air, there is no mechanism for setting up the backward wave $\mathscr{E}_3^-$ so we put this equal to zero. Using these two facts then together with the chain rule for matrices we conclude:

$$\begin{bmatrix} \mathscr{E}_{s1}^+ \\ \mathscr{E}_{s1}^- \end{bmatrix} = \begin{bmatrix} \dfrac{\eta+\eta_0}{2\eta} & \dfrac{\eta-\eta_0}{2\eta} \\ \dfrac{\eta-\eta_0}{2\eta} & \dfrac{\eta+\eta_0}{2\eta} \end{bmatrix} \begin{bmatrix} e^{j\beta l} & 0 \\ 0 & e^{-j\beta l} \end{bmatrix} \begin{bmatrix} \dfrac{\eta_0+\eta}{2\eta_0} & \dfrac{\eta_0-\eta}{2\eta_0} \\ \dfrac{\eta_0-\eta}{2\eta_0} & \dfrac{\eta_0+\eta}{2\eta_0} \end{bmatrix} \begin{bmatrix} \mathscr{E}_{r3}^+ \\ 0 \end{bmatrix}$$

$$= \begin{bmatrix} A_{11} & A_{12} \\ A_{21} & A_{22} \end{bmatrix} \begin{bmatrix} \mathscr{E}_{r3}^+ \\ 0 \end{bmatrix}$$

Now $\mathscr{E}_{s1}^- = A_{21}\mathscr{E}_{r3}^+$ and $\mathscr{E}_{s1}^+ = A_{11}\mathscr{E}_{r3}^+$

Therefore the reflection coefficient in the air at the slab face is

$$\frac{\mathscr{E}_{s1}^-}{\mathscr{E}_{s1}^+} = \frac{A_{21}}{A_{11}} = \frac{\left(\dfrac{\eta-\eta_0}{2\eta}\right) e^{j\beta l}\left(\dfrac{\eta_0+\eta}{2\eta_0}\right) + \left(\dfrac{\eta+\eta_0}{2\eta}\right) e^{-j\beta l}\left(\dfrac{\eta_0-\eta}{2\eta_0}\right)}{\left(\dfrac{\eta+\eta_0}{2\eta}\right) e^{j\beta l}\left(\dfrac{\eta_0+\eta}{2\eta_0}\right) + \left(\dfrac{\eta-\eta_0}{2\eta}\right) e^{-j\beta l}\left(\dfrac{\eta_0-\eta}{2\eta_0}\right)}$$

$$= \frac{(\eta^2-\eta_0^2)\, 2\, j \sin\beta l}{(\eta^2+2\eta\eta_0+\eta_0^2)\, e^{j\beta l} - (\eta^2-2\eta\eta_0+\eta_0^2)\, e^{-j\beta l}}$$

$$= \frac{(\eta^2-\eta_0^2)\, j \sin\beta l}{(\eta^2+\eta_0^2)\, j \sin\beta l + 2\eta\eta_0 \cos\beta l}$$

$$= \left(\frac{\eta^2-\eta_0^2}{\eta^2+\eta_0^2}\right) \frac{1}{\left(1 - j\dfrac{2\eta\eta_0}{(\eta^2+\eta_0^2)} \cot\beta l\right)}$$

η and η_0 are both real numbers since both media are lossless.

4.7. Proceeding as we did in Problem **4.6** we see that

$$\frac{\mathscr{E}_{r3}^+}{\mathscr{E}_{s1}^+} = \frac{1}{A_{11}}$$

In this case γ and η are both complex in general.

Now
$$\frac{1}{A_{11}} = \frac{1}{\dfrac{(\eta^2 + 2\eta\eta_0 + \eta_0^2)}{4\eta\eta_0} e^{\gamma l} - \dfrac{(\eta^2 - 2\eta\eta_0 + \eta_0^2)}{4\eta\eta_0} e^{-\gamma l}}$$

$$= \frac{2\eta\eta_0}{(\eta^2 + \eta_0^2) \sinh \gamma l + 2\eta\eta_0 \cosh \gamma l}$$

i.e.
$$\frac{\mathscr{E}^+_{r3}}{\mathscr{E}^+_{s1}} = \frac{2\eta\eta_0}{(\eta^2 + \eta_0^2)} \cdot \frac{1}{\sinh \gamma l + \dfrac{2\eta\eta_0}{(\eta^2 + \eta_0^2)} \cosh \gamma l}$$

To answer the numerical part of this question we use some rather well-known results:

$$\gamma = \alpha + j\beta$$

where
$$\alpha = \frac{2\pi}{\lambda} [\tfrac{1}{2}(\sqrt{1 + \tan^2 \delta} - 1)]^{1/2}$$

and
$$\beta = \frac{2\pi}{\lambda} [\tfrac{1}{2}(\sqrt{1 + \tan^2 \delta} + 1)]^{1/2}$$

also
$$\eta = \sqrt{\frac{j\omega\mu_0}{\sigma + j\omega\epsilon}} = \frac{\eta_0}{\sqrt{\epsilon_r}\sqrt{1 - j \tan \delta}}$$

In our particular case

$$\alpha = \frac{2\pi}{\lambda} [\tfrac{1}{2}(\sqrt{1 + 0.0004} - 1)]^{1/2} \doteqdot \frac{2\pi}{\lambda} \cdot \frac{1}{100}$$

$$\beta \doteqdot \frac{2\pi}{\lambda}$$

$\therefore$
$$\beta l \doteqdot \pi$$

and
$$\alpha l \approx \frac{\pi}{100}$$

$$\frac{\eta_0}{\eta} = 3.16\sqrt{1 - j\,0.02} \doteqdot 3.16\,(1 - j\,0.01)$$

Now $\dfrac{2\eta\eta_0}{\eta^2+\eta_0{}^2} = \dfrac{2(\eta_0/\eta)}{1+(\eta_0/\eta)^2} = \dfrac{6\cdot32\,(1-\mathrm{j}\,0\cdot01)}{11-\mathrm{j}\,0\cdot2} \approx 0\cdot575$

Now $\cosh(\alpha l + \mathrm{j}\pi) = -\cosh\alpha l \approx -1$ for small arguments,
and $\sinh(\alpha l + \mathrm{j}\pi) = -\sinh\alpha l \approx -\alpha l$ for small arguments.
Hence

$$\frac{\mathscr{E}^+_{r3}}{\mathscr{E}^+_{s1}} \doteqdot \frac{0\cdot575}{-0\cdot0314 - 0\cdot575} = -0\cdot948$$

Thus the efficiency of transmission $= \left|\dfrac{\mathscr{E}^+_{r3}}{\mathscr{E}^+_{s1}}\right|^2 \times 100\%$

$\doteqdot 90\%$

4.8. The total field at r is the sum of the forward and backward waves, i.e.

$$E_r = \mathscr{E}^+_{r2} + \mathscr{E}^-_{r2} \qquad \ldots.(1)$$

and

$$H_r = \mathscr{H}^+_{r2} - \mathscr{H}^-_{r2} \qquad \ldots.(2)$$

$\therefore$

$$\eta H_r = \mathscr{E}^+_{r2} - \mathscr{E}^-_{r2} \qquad \ldots.(3)$$

From Equations (1) and (3)

$$\mathscr{E}^+_{r2} = \tfrac{1}{2}[E_r + \eta H_r] \qquad \ldots.(4)$$

$$\mathscr{E}^-_{r2} = \tfrac{1}{2}[E_r - \eta H_r] \qquad \ldots.(5)$$

Now $E_s = \mathscr{E}^+_{s2} + \mathscr{E}^-_{s2} = \mathscr{E}^+_{r2}\,\mathrm{e}^{\gamma l} + \mathscr{E}^-_{r2}\,\mathrm{e}^{-\gamma l}$

and $H_s = \mathscr{H}^+_{s2} - \mathscr{H}^-_{s2} = \mathscr{H}^+_{r2}\,\mathrm{e}^{\gamma l} - \mathscr{H}^-_{r2}\,\mathrm{e}^{-\gamma l}$ $\qquad \ldots.(6)$

Substituting Equations (4) and (5) into Equation (6)

$$\left.\begin{aligned} E_s &= \tfrac{1}{2}[E_r + \eta H_r]\,\mathrm{e}^{\gamma l} + \tfrac{1}{2}[E_r - \eta H_r]\,\mathrm{e}^{-\gamma l} = E_r \cosh\gamma l + H_r\eta \sinh\gamma l \\ H_s &= \frac{1}{2\eta}[E_r + \eta H_r]\,\mathrm{e}^{\gamma l} - \frac{1}{2\eta}[E_r - \eta H_r]\,\mathrm{e}^{-\gamma l} = E_r\frac{\sinh\gamma l}{\eta} + H_r\cosh\gamma l \end{aligned}\right\} \ldots(7)$$

In matrix notation, Equation (7) becomes

$$\begin{bmatrix} E_s \\ H_s \end{bmatrix} = \begin{bmatrix} (\cosh \gamma l) & (\eta \sinh \gamma l) \\ \left(\frac{1}{\eta} \cdot \sinh \gamma l\right) & (\cosh \gamma l) \end{bmatrix} \begin{bmatrix} E_r \\ H_r \end{bmatrix} \qquad \ldots . (8)$$

From Equation (7):

$$\frac{E_s}{H_s} = \frac{(E_r/H_r) \cosh \gamma l + \eta \sinh \gamma l}{1/\eta \, (E_r/H_r) \sinh \gamma l + \cosh \gamma l} \qquad \ldots . (9)$$

Defining the ratio E/H as a 'wave impedance', Z, we obtain on rearranging Equation (9)

$$Z_s = \eta \left(\frac{Z_r + \eta \tanh \gamma l}{\eta + Z_r \tanh \gamma l} \right) \qquad \ldots . (10)$$

which for a lossless media becomes

$$Z_s = \eta \left(\frac{Z_r + \mathrm{j}\, \eta \tan \beta l}{\eta + \mathrm{j}\, Z_r \tan \beta l} \right) \qquad \ldots . (11)$$

η, the intrinsic impedance of the medium, is now real and β is the imaginary part of the propagation constant, γ.

If we replace E by V and H by I in Equation (8), where V and I refer to voltage and current respectively, we see that the matrix in this equation is the *ABCD* matrix of a uniform transmission line having a characteristic impedance η.

Equations (10) and (11) are the well known 'impedance transformation' equations of uniform transmission line theory and so we are at liberty to use this theory in the solution of plane wave problems. In particular, the use of the Smith Chart for numerical problems is very convenient. It also affords a check on analytical solutions.

4.9. *Figure 4.7* shows the equivalent transmission line problem where we have replaced the semi-infinite region on the far side of the slab by its characteristic impedance η_0.

We now normalize the load impedance η_0 to the line impedance, $\eta_0/3$, i.e.

$$z_l = \frac{\eta_0}{\eta_0/3} = 3 + \mathrm{j}\,0$$

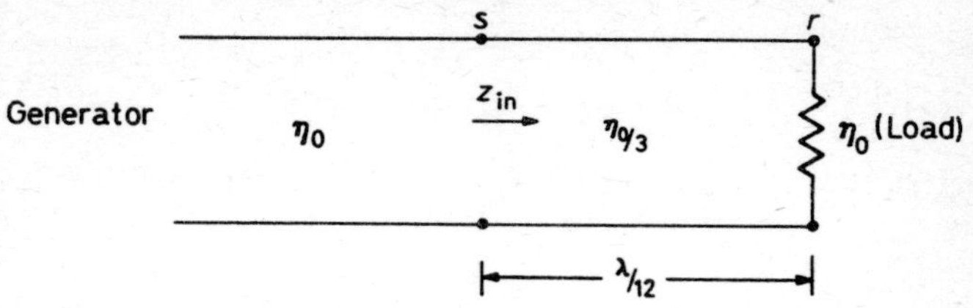

Figure 4.7.

Enter Smith Chart, *Figure 4.8*, at $z_l = 3 + j\,0$.
Move around circle of constant reflection coefficient a distance 0·0834 λ towards the generator (clockwise) and read off the normalized input impedance, z_{in}, i.e.

$$z_{in} = \frac{Z_{in}}{\eta_0/3} = 1 - j1{\cdot}15$$

$$\therefore \qquad Z_{in} = \eta_0(0{\cdot}34 - j0{\cdot}383)$$

Normalize Z_{in} to the first line, i.e.

$$\frac{Z_{in}}{\eta_0} = 0{\cdot}34 - j0{\cdot}383$$

Enter chart at $0{\cdot}34 - j0{\cdot}383$.
Therefore, magnitude of the reflection coefficient on first line

$$= \frac{\text{length of } OP}{\text{radius of chart}}$$

$$= \frac{4{\cdot}6}{8{\cdot}3} = 0{\cdot}555$$

Angle of reflection coefficient at plane $s = 225°$, read from chart.

Let us now check our answers by using the equation developed in Problem **4.6.**

$$\frac{\eta^2 - \eta_0^2}{\eta^2 + \eta_0^2} = \frac{1/9 - 1}{1/9 + 1} = -0{\cdot}8$$

$$\frac{2\eta\eta_0 \cot \beta l}{\eta^2 + \eta_0^2} = \frac{2(1/3)\sqrt{3}}{1/9 + 1} = 0{\cdot}6\sqrt{3}$$

Therefore, angle of reflection coefficient

$$= \pi + \tan^{-1}(0{\cdot}6\sqrt{3}) = 226{\cdot}1°$$

and magnitude of reflection coefficient

$$= \frac{0{\cdot}8}{\sqrt{1 + 1{\cdot}08}} = 0{\cdot}555$$

We see that the agreement between the two methods is good. From a practical point of view there is little to go wrong with the Smith Chart approach, which merely involves the plotting of a few complex numbers.

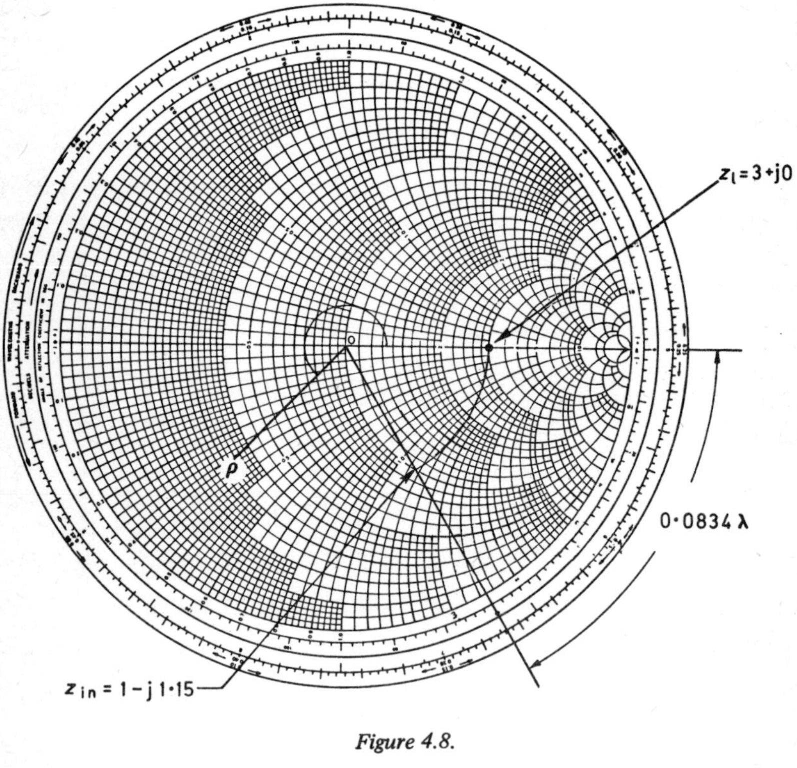

Figure 4.8.

4.10. A little thought will convince the reader that we can write the solution of this problem from the result of Problem **4.6**, the modification being slight. We will not do this however but will first develop some rather general results which can be applied to a large class of problems. From these we will then prove that our equation is valid.

Consider first the exponential factors associated with the forward and backward waves in the dielectric. Because the dielectric is lossless and because also of our choice of coordinates, these become

$$e^{j(\omega t - \beta x \sin\phi - \beta z \cos\phi)} \qquad \text{(forward wave)}$$

$$e^{j(\omega t - \beta x \sin\phi + \beta z \cos\phi)} \qquad \text{(backward wave)}$$

The point to note here is that in the direction normal to the interface, i.e. z, the propagation constant is now $\beta \cos\phi$. For convenience we will again suppress the $e^{j\omega t}$. Hence

$$\begin{bmatrix} \mathscr{E}^+_{s2} \\ \mathscr{E}^-_{s2} \end{bmatrix} = \begin{bmatrix} e^{j\beta d\cos\phi} & 0 \\ 0 & e^{-j\beta d\cos\phi} \end{bmatrix} \begin{bmatrix} \mathscr{E}^+_{r2} \\ \mathscr{E}^-_{r2} \end{bmatrix} \qquad \ldots\ldots(1)$$

As in Problem **4.5** we will also determine the wave amplitude matrix for the junction s by equating the tangential components of the field across this junction, i.e.

$$\left.\begin{aligned} (\mathscr{E}^+_{s1} + \mathscr{E}^-_{s1})\, e^{-j\beta_0 x \sin\theta} &= (\mathscr{E}^+_{s2} + \mathscr{E}^-_{s2})\, e^{-j\beta x \sin\phi} \\ (\mathscr{H}^+_{s1} - \mathscr{H}^-_{s1}) \frac{e^{-j\beta_0 x \sin\theta}}{\sec\theta} &= (\mathscr{H}^+_{s2} - \mathscr{H}^-_{s2}) \frac{e^{-j\beta x \sin\phi}}{\sec\phi} \end{aligned}\right\} \qquad \ldots\ldots(2)$$

These relations (Equations 2) must be true for all x.

$$\therefore \qquad \beta_0 \sin\theta = \beta \sin\phi \qquad \text{(Snell's law)} \ldots (3)$$

Using the facts that

$$\frac{\mathscr{E}^+_{s1}}{\mathscr{H}^+_{s1}} = \frac{\mathscr{E}^-_{s1}}{\mathscr{H}^-_{s1}} = \eta_0$$

and

$$\frac{\mathscr{E}^+_{s2}}{\mathscr{H}^+_{s2}} = \frac{\mathscr{E}^-_{s2}}{\mathscr{H}^-_{s2}} = \eta$$

in Equation (2) we obtain the matrix for the junction:

$$\begin{bmatrix} \mathscr{E}_{s1}^{+} \\ \mathscr{E}_{s1}^{-} \end{bmatrix} = \begin{bmatrix} \dfrac{\eta \sec\phi + \eta_0 \sec\theta}{2\eta \sec\phi} & \dfrac{\eta \sec\phi - \eta_0 \sec\theta}{2\eta \sec\phi} \\ \dfrac{\eta \sec\phi - \eta_0 \sec\theta}{2\eta \sec\phi} & \dfrac{\eta \sec\phi + \eta_0 \sec\theta}{2\eta \sec\phi} \end{bmatrix} \begin{bmatrix} \mathscr{E}_{s2}^{+} \\ \mathscr{E}_{s2}^{-} \end{bmatrix} \quad \ldots. (4)$$

From Equations (1) and (4) it is clear that we can use the solution to Problem 4.6 provided that we replace η by $\eta \sec\phi$
η_0 by $\eta_0 \sec\theta$
β by $\beta \cos\phi$
where θ and ϕ are related by Snell's law, Equation (3).

$$\therefore \qquad \Gamma = \frac{\eta^2 \sec^2\phi - \eta_0{}^2 \sec^2\theta}{(\eta^2 \sec^2\phi + \eta_0{}^2 \sec^2\theta) - \mathrm{j}2\eta\eta_0 \sec\phi \sec\theta \cot\{\beta d \cos\phi\}}$$

Now $\eta = \eta_0/\sqrt{K}$

$$\therefore \qquad \eta_0 = \sqrt{K}\,\eta$$

and since θ is given as 45°, $\sec\theta = \sqrt{2}$

$$\therefore \qquad \Gamma = \frac{1 - 2K\cos^2\phi}{(1 + 2K\cos^2\phi) - \mathrm{j}\,2\sqrt{2K}\cos\phi \cot\{\beta d \cos\phi\}}$$

From Snell's law,

$$\frac{\sin\theta}{\sin\phi} = \sqrt{K}$$

$$\therefore \qquad \sin\phi = \frac{\sin\theta}{\sqrt{K}} = \frac{1}{\sqrt{2K}}$$

hence $$\cos\phi = \frac{\sqrt{2K - 1}}{\sqrt{2K}}$$

and
$$\Gamma = \frac{-(K-1)}{K - j\sqrt{2K-1}\cot\{(2\pi d/\lambda_0)\sqrt{K-½}\}}$$

Now
$$|\Gamma|^2 = \frac{(K-1)^2}{K^2 + (2K-1)\cot^2\psi} = \frac{1}{2}$$

where
$$\psi = \frac{2\pi d}{\lambda_0}\sqrt{K-½}$$

Putting
$$K = 5{\cdot}4 \text{ gives } \cot\psi = 0{\cdot}995$$

$\therefore$
$$\psi = 2\pi\frac{d}{\lambda_0}2{\cdot}216 \doteqdot \frac{\pi}{4}$$

$\therefore$
$$d = 0{\cdot}508 \text{ mm } (0{\cdot}020 \text{ in.})$$

4.11. Making use of the wave amplitude matrices given in previous problems we can write the matrix for this one by inspection:

$$\begin{bmatrix} \mathscr{E}^+_{s1} \\ \mathscr{E}^-_{s1} \end{bmatrix} = \begin{bmatrix} a & b \\ b & a \end{bmatrix}\begin{bmatrix} e^{j\beta_0 d\cos\phi} & 0 \\ 0 & e^{-j\beta_0 d\cos\phi} \end{bmatrix}\begin{bmatrix} f & g \\ g & f \end{bmatrix}\begin{bmatrix} \mathscr{E}^+_{r3} \\ 0 \end{bmatrix} = \begin{bmatrix} A_{11} & A_{12} \\ A_{21} & A_{22} \end{bmatrix}\begin{bmatrix} \mathscr{E}^+_{r3} \\ 0 \end{bmatrix}$$

where
$$a = \frac{\eta_0\sec\phi + \eta\sec\theta}{2\eta_0\sec\phi},$$

$$b = \frac{\eta_0\sec\phi - \eta\sec\theta}{2\eta_0\sec\phi}.$$

$$f = \frac{\eta\sec\theta + \eta_0\sec\phi}{2\eta\sec\theta}.$$

$$g = \frac{\eta\sec\theta - \eta_0\sec\phi}{2\eta\sec\theta}$$

and θ and ϕ are related by Snell's law, i.e.

$$\sin\phi = \sqrt{K}\sin\theta$$

$\therefore$
$$\cos\phi = \sqrt{1 - K\sin^2\theta} = j\sqrt{K\sin^2\theta - 1} \qquad \ldots .(1)$$

We must now apply a little physical reasoning to see what sign we give to $\cos\phi$ since the square root sign yields + or −.

Since $\cos\phi$ is imaginary, $\theta > \theta_c$, the exponential terms become real, inferring exponential decays and growths with distance. From physical reasoning we wish to associate the growth term with the factor

$$e^{j\beta_0 d \cos\phi}$$

$$\therefore \qquad \cos\phi = -j \times (\text{some real term})$$

This can easily be satisfied if we replace Equation (1) by

$$\cos\phi = -j\sqrt{K \sin^2\theta - 1}$$

and consider only the positive root.

Now $\quad \mathscr{E}^+_{s1} = A_{11}\,\mathscr{E}^+_{r3}$

$$\therefore \qquad \frac{\mathscr{E}^+_{s1}}{\mathscr{E}^+_{r3}} = A_{11} = af\,e^{\alpha d} + bg\,e^{-\alpha d}, \qquad \alpha \equiv j\beta_0 \cos\phi$$

$$= \frac{(\eta_0 \sec\phi + \eta \sec\theta)^2\, e^{\alpha d} - (\eta_0 \sec\phi - \eta \sec\theta)^2\, e^{-\alpha d}}{4\eta_0\eta \sec\phi \sec\theta}$$

$$= \cosh \alpha d + \frac{\eta_0^2 \sec^2\phi + \eta^2 \sec^2\theta}{2\eta_0\eta \sec\phi \sec\theta} \sinh \alpha d$$

$$= \cosh \alpha d + j\left[\frac{K(\cos^2\theta - \sin^2\theta) + 1}{2\sqrt{K}\cos\theta\sqrt{K\sin^2\theta - 1}}\right] \sinh \alpha d$$

$$\therefore \left|\frac{\mathscr{E}^{+}_{s1}}{\mathscr{E}^{+}_{r3}}\right|^2 = \left|\frac{\mathscr{E}^{+}_{1}}{\mathscr{E}^{+}_{3}}\right|^2$$

$$= \cosh^2 \alpha d + \left(\frac{K^2 [\cos^4\theta - 2\cos^2\theta \sin^2\theta + \sin^4\theta] + 2K(\cos^2\theta - \sin^2\theta) + 1}{4K\cos^2\theta\,(K\sin^2\theta - 1)}\right) \times \sinh^2 \alpha d$$

$$= \cosh^2 \alpha d + \left(\frac{K^2(\cos^2\theta + \sin^2\theta)^2 - 2K + 1 - 4K\cos^2\theta\,(K\sin^2\theta - 1)}{4K\cos^2\theta\,(K\sin^2\theta - 1)}\right)\sinh^2 \alpha d$$

$$\therefore \qquad \left|\frac{\mathscr{E}^{+}_{1}}{\mathscr{E}^{+}_{3}}\right|^2 = 1 + \frac{(K-1)^2 \sinh^2 \alpha d}{4K\cos^2\theta\,(K\sin^2\theta - 1)}$$

Note. For an interesting account of the practical application of the principles involved in this and the previous problem, the reader should consult: 'Millimetre Wave Transmission by Non-waveguide Means' by Rufus G. Fellers, *Microwave Journal*, 1962, **5** (May), 80–86.

CHAPTER 5

5.1. From the symmetry of the problem we see that the current density and fields will vary only with the radius, r, i.e.

$$\frac{\partial}{\partial z} = \frac{\partial}{\partial \phi} = 0$$

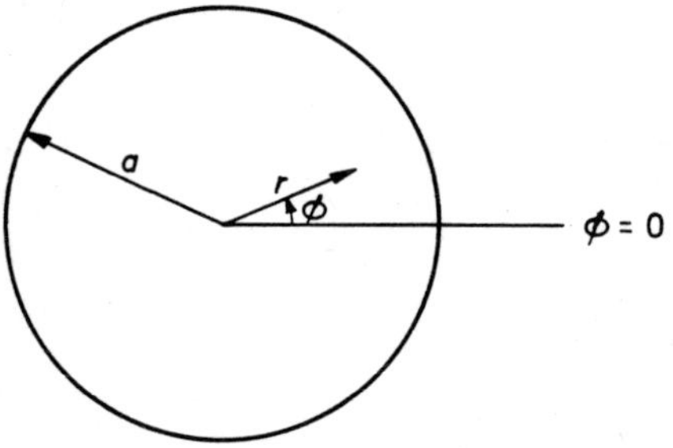

Figure 5.3.

The current density vector is in the z direction and **H** has only an H_ϕ component (see *Figure 5.3*). Using these facts, Maxwell's curl equations reduce to:

$$\frac{1}{r}\frac{\partial}{\partial r}(rH_\phi) = \sigma E_z \qquad \ldots .(1)$$

$$\frac{\partial E_z}{\partial r} = \mathrm{j}\omega\mu H_\phi \qquad \ldots .(2)$$

where we have ignored the displacement current within the conducting wire and have assumed an exponential time dependence, $e^{j\omega t}$.

σ and μ are the conductivity and permeability of the wire.

At zero frequency, we see from Equation (2) that the current density is uniform, for instance σE_0. We can now use this in Equation (1) to give us the zero frequency magnetic field, H_0, i.e.

$$\frac{1}{r}\frac{\partial}{\partial r}(rH_0) = \sigma E_0$$

$$\therefore \qquad H_0 = \frac{\sigma E_0}{2} r$$

Substituting H_0 into Equation (2) will yield the first order correction in E, E_1, i.e.

$$\frac{\partial E_1}{\partial r} = \mathrm{j}\,\frac{\omega\mu\sigma}{2} E_0\, r$$

$$\therefore \qquad E_1 = \mathrm{j}\,\frac{\omega\mu\sigma E_0}{2}\,\frac{r^2}{2}$$

Substituting E_1 into Equation (1) gives us the first order correction in H, H_1.

$$\frac{1}{r}\frac{\partial}{\partial r}(rH_1) = \mathrm{j}\,\frac{\omega\mu\sigma^2 E_0}{4} r^2$$

$$\therefore \qquad H_1 = \mathrm{j}\,\frac{\omega\mu\sigma^2 E_0}{16} r^3$$

Continuing the process

$$\frac{\partial E_2}{\partial r} = \frac{(\mathrm{j}\omega\mu\sigma)^2 E_0 r^3}{16}$$

$$\therefore \qquad E_2 = -\left(\frac{\omega\mu\sigma}{2}\right)^2 \frac{E_0}{4}\,\frac{r^4}{4}$$

At this point it may be a good idea to stop and consider where all this is leading us to. What we wish to do is to calculate the power loss in the conductor at some frequency and compare this with that loss which would be produced if the same total current were distributed uniformly throughout the cross section of the conductor (as it would be at zero frequency). Since the first correction term in E is to the second power of (r/δ), then the first correction in the loss will be to $(r/\delta)^4$.

Here δ is the skin depth factor

$$\left(\frac{2}{\omega\mu\sigma}\right)^{1/2}$$

All we need do then is to ensure that we have all field terms to the order $(r/\delta)^2$ which we have already got.

Therefore the actual power loss, $P_{\text{a.c.}} = \dfrac{\sigma}{2}\displaystyle\int_V \mathbf{E}\,.\,\mathbf{E}^*\,\mathrm{d}V$ per unit length

$$= \sigma\pi E_0{}^2 \int_0^a \left(1 + \mathrm{j}\frac{r^2}{2\delta^2} - \frac{1}{16}\frac{r^4}{\delta^4}\right)\left(1 - \mathrm{j}\frac{r^2}{2\delta^2} - \frac{1}{16}\frac{r^4}{\delta^4}\right) r\,\mathrm{d}r$$

$$= \sigma\pi E_0{}^2 \int_0^a \left(r + \frac{1}{8}\frac{r^5}{\delta^4}\right)\mathrm{d}r$$

$$\therefore \quad P_{\text{a.c.}} \doteqdot \frac{\sigma\pi E_0{}^2 a^2}{2}\left[1 + \frac{1}{24}\frac{a^4}{\delta^4}\right] \qquad \ldots.(3)$$

Now the total current flowing in the wire is simply

$$I \doteqdot \sigma E_0 \int_0^a \left(1 + \mathrm{j}\frac{r^2}{2\delta^2} - \frac{1}{16}\frac{r^4}{\delta^4} + \ldots\right) 2\pi r\,\mathrm{d}r$$

and if this current were to flow through the d.c. resistance of the wire, the loss per unit length would be

$$P_{\text{d.c.}} = \tfrac{1}{2}\, I\, I^* \frac{1}{\sigma\pi a^2}$$

Now $$I = \sigma E_0 \pi a^2 \left[\left(1 - \frac{1}{48}\frac{a^4}{\delta^4}\right) + \mathrm{j}\frac{1}{4}\frac{a^2}{\delta^2}\right]$$

$$\therefore \quad P_{\text{d.c.}} = \frac{1}{2}\left[\sigma^2 E_0{}^2 \pi^2 a^4 \left(1 - \frac{2}{48}\frac{a^4}{\delta^4} + \frac{1}{16}\frac{a^4}{\delta^4}\right)\right]\frac{1}{\sigma\pi a^2}$$

$$\therefore \quad P_{\text{d.c.}} = \frac{\sigma\pi E_0{}^2 a^2}{2}\left(1 + \frac{1}{48}\frac{a^4}{\delta^4}\right) \qquad \ldots.(4)$$

hence
$$\frac{P_{\text{a.c.}}}{P_{\text{d.c.}}} = \frac{R_{\text{a.c.}}}{R_{\text{d.c.}}} = \left(1 + \frac{1}{24}\frac{a^4}{\delta^4}\right)\left(1 + \frac{1}{48}\frac{a^4}{\delta^4}\right)^{-1}$$

$$\therefore \qquad \frac{R_{\text{a.c.}}}{R_{\text{d.c.}}} = \left(1 + \frac{1}{48}\frac{a^4}{\delta^4}\right)$$

As an alternative approach to this problem let us form $2\pi a\,(\mathbf{E} \times \mathbf{H}^*)$. By inspection of the terms of this expression, together with a knowledge of the complex Poynting theorem we should be able to determine the correction factor we are looking for. Now at $r = a$,

$$E = E_0\left[1 + \text{j}\frac{1}{2}\frac{a^2}{\delta^2} - \frac{a^4}{16\delta^4} + \text{j}\,0\left[\frac{a^6}{\delta^6}\right]\right]$$

$$H = \frac{\sigma E_0 a}{2}\left[1 + \text{j}\frac{1}{4}\cdot\frac{a^2}{\delta^2} - \frac{1}{48}\frac{a^4}{\delta^4} + \text{j}\,0\left[\frac{a^6}{\delta^6}\right]\right]$$

where we have calculated a further term in H to bring it up to the fourth power in a/δ. Hence

$$2\pi a\,(\mathbf{E} \times \mathbf{H}^*) = (\sigma E_0^2 \pi a^2)\left[1 + \frac{2}{48}\frac{a^4}{\delta^4} + \text{j}\frac{1}{4}\cdot\frac{a^2}{\delta^2}\right]$$

to this degree of approximation.

Now
$$\tfrac{1}{2}\oint \mathbf{E} \times \mathbf{H}^* \cdot d\mathbf{S} = \tfrac{1}{2} II^* (R_{\text{a.c.}} + \text{j}\omega L)$$

$$\therefore \qquad R_{\text{a.c.}} \doteqdot \frac{\sigma E_0^2 \pi a^2\left[1 + \frac{2}{48}\frac{a^4}{\delta^4}\right]}{II^*} = \frac{1}{\sigma\pi a^2}\,\frac{\left[1 + \frac{2}{48}\frac{a^4}{\delta^4}\right]}{\left[1 + \frac{1}{48}\frac{a^4}{\delta^4}\right]}$$

Whence, to order of a^4/δ^4,

$$\frac{R_{\text{a.c.}}}{R_{\text{d.c.}}} = 1 + \frac{1}{48}\frac{a^4}{\delta^4}$$

In a similar manner, $L = \mu/8\pi$ to first order.

5.2. Assuming the iron is unsaturated, its relative permeability μ_r will be of the order 2,000 to 3,000 or more whilst that of the copper bar will be 1. Hence we can assume that the **H** field will only have a y component in the slot and will be zero in the iron (see *Figure 5.1,* page 31.

Ignoring the displacement current in the copper and assuming a time variation of the form $e^{j\omega t}$, we obtain from Maxwell's equations

$$\frac{\partial H_y}{\partial x} = \sigma E_z \quad \ldots.(1)$$

$$\frac{\partial E_z}{\partial x} = j\omega\mu H_y \quad \ldots.(2)$$

If the field were stationary, the electric field would be uniform and of value E_0 say. Equations (1) and (2) then reduce to:

$$\frac{\partial H_0}{\partial x} = \sigma E_0 \quad \ldots.(3)$$

$$\frac{\partial E_0}{\partial x} = 0 \quad \ldots.(4)$$

Thus the zero order magnetic field is $H_0 = x\sigma E_0$

From Equation (2) we can get the first order correction in the electric field.

$$\frac{\partial E_1}{\partial x} = j\omega\mu\sigma E_0 x$$

$$\therefore \quad E_1 = j\left(\frac{\omega\mu\sigma}{2}\right) E_0 x^2$$

The first order correction in H is

$$\frac{\partial H_1}{\partial x} = j\left(\frac{\omega\mu\sigma}{2}\right) E_0 \sigma x^2$$

$$\therefore \quad H_1 = j\left(\frac{\omega\mu\sigma}{2}\right) \sigma E_0 \frac{x^3}{3}$$

And finally, the second order correction in E is

$$\frac{\partial E_2}{\partial x} = \frac{(j\omega\mu\sigma)^2}{2} E_0 \frac{x^3}{3}$$

$$\therefore \quad E_2 = -\left(\frac{\omega\mu\sigma}{2}\right)^2 \frac{E_0 x^4}{6}$$

$$\therefore \quad E \doteqdot E_0 + E_1 + E_2$$

which in terms of the skin depth factor δ becomes

$$E = E_0 \left[1 + j\frac{x^2}{\delta^2} - \frac{x^4}{6\delta^4} \right] \qquad \ldots .(5)$$

$$\text{Therefore total current, } I = \sigma E_0 W \int_0^h \left(1 + j\frac{x^2}{\delta^2} - \frac{x^4}{6\delta^4}\right) dx$$

$$= \sigma E_0 Wh \left[\left(1 - \frac{s^4}{30}\right) + j\frac{s^2}{3} \right]$$

If this current flowed through the d.c. resistance of the conductor the loss would be:

$$P_{d.c.} \doteqdot \frac{1}{2}\left[\sigma^2 W^2 E_0{}^2 h^2 \left(1 + \frac{2s^4}{45}\right)\right] \frac{1}{\sigma Wh}$$

$$\therefore \qquad P_{d.c.} \doteqdot \frac{Wh\sigma E_0{}^2}{2}\left(1 + \frac{2s^4}{45}\right) \qquad \ldots .(6)$$

$$\text{Now} \qquad P_{a.c.} = \frac{W\sigma}{2} \int_0^h \mathbf{E} \,.\, \mathbf{E}^* \, dx$$

$$\doteqdot \frac{W\sigma E_0{}^2}{2} \int_0^h \left[\left(1 - \frac{x^4}{6\delta^4}\right)^2 + \frac{x^4}{\delta^4} \right] dx$$

$$\doteqdot \frac{W\sigma E_0{}^2}{2} \int_0^h \left(1 + \frac{2}{3}\frac{x^4}{\delta^4}\right) dx$$

$$\therefore \qquad P_{a.c.} \doteqdot \frac{W\sigma h E_0{}^2}{2}\left(1 + \frac{2s^4}{15}\right) \qquad \ldots .(7)$$

From Equations (6) and (7)

$$\frac{P_{\text{a.c.}}}{P_{\text{d.c.}}} = \frac{R_{\text{a.c.}}}{R_{\text{d.c.}}} = \left(1 + \frac{2s^4}{15}\right)\left(1 + \frac{2s^4}{45}\right)^{-1}$$

$$\therefore \qquad \frac{R_{\text{a.c.}}}{R_{\text{d.c.}}} \doteqdot \left(1 + \frac{4s^4}{45}\right) \qquad \ldots.(8)$$

5.3. Making use of the symmetry of the problem and the fact that the coil is long, we shall assume that the magnetic field has only a longitudinal component which can vary with radius and time. Thus ignoring the displacement current in the rod and assuming an $e^{j\omega t}$ time dependence, Maxwell's equations reduce to:

$$\frac{\partial H_z}{\partial r} = -\sigma E_\theta \qquad \ldots.(1)$$

$$\frac{1}{r}\frac{\partial}{\partial r}(rE_\theta) = -j\omega\mu H_z \qquad \ldots.(2)$$

If the magnetic field H_0, say, was steady, then the electric field in the brass rod would be zero. The above equations then give

$$\frac{\partial H_0}{\partial r} = 0 \qquad \ldots.(3)$$

$$\frac{1}{r}\frac{\partial}{\partial r}(rE_\theta) = -j\omega\mu H_0 \qquad \ldots.(4)$$

Integrating Equation (4) will now give us the first order correction in E, say E_1, i.e.

$$E_1 = -j\frac{\omega\mu H_0}{2}r$$

Now this electric field must be continuous across the brass rod interface with the coil insulation and since this is thin, E_1 must be induced in the coil. But the coil is said to have zero resistance and therefore cannot support the field E_1, which must be balanced by the source producing the current flow; i.e.

Voltage of source = $\left\{ \mathrm{j}\dfrac{\omega\mu H_0}{2} a \right\}$ integrated along the length of wire in the coil

$$\therefore \qquad V = \mathrm{j}\frac{\omega\mu H_0 a}{2} \cdot 2\pi a N$$

But $$H_0 = \frac{NI}{l} \quad \text{(where } I \text{ is the current in the coil)}$$

$$\therefore \qquad V = \mathrm{j}\omega\left(\pi a^2 \mu \frac{N^2}{l}\right) I$$

$$\therefore \qquad \frac{V}{I} = Z = \mathrm{j}\omega L \qquad \ldots . (5)$$

where L we recognise as simply the inductance of the coil calculated by magnetostatic methods.

To the first approximation then the coil is simply an inductor whose inductance is the same as that calculated at d.c.

Let us now put E_1 into Equation (1) to obtain the first correction in H; i.e.

$$\frac{\partial H_1}{\partial r} = -\sigma\left(-\mathrm{j}\frac{\omega\mu H_0}{2}\right) r$$

$$\therefore \qquad H_1 = \mathrm{j}\frac{\omega\mu\sigma}{2} H_0 \frac{r^2}{2} + \text{constant}$$

If the current in the coil is the same as before, then the H at the rod surface must remain the same; i.e. at $r = a$, $H = H_0$. Hence

$$H_1 = \mathrm{j}\frac{\omega\mu\sigma}{4} H_0 (r^2 - a^2)$$

This in turn gives rise to an electric field contribution given by

$$\frac{1}{r}\frac{\partial}{\partial r}(rE_2) = -\mathrm{j}\omega\mu\left(\mathrm{j}\frac{\omega\mu\sigma}{4} H_0\right)(r^2 - a^2)$$

$$\therefore \qquad rE_2 = \frac{\omega^2\mu^2\sigma H_0}{4}\left(\frac{r^4}{4} - a^2\frac{r^2}{2}\right)$$

Thus at $r = a$,

$$E_2 = -\frac{\omega^2\mu^2\sigma H_0 a^3}{16}$$

The total applied voltage to the coil is now

$$V = -(E_1 + E_2)\, 2\pi a N$$

$$\therefore \quad \frac{V}{I} = j\omega L + \frac{\omega^2\mu^2\sigma}{16}\frac{N^2}{l} \,.\, 2\pi a^4$$

$$= \omega L\left[j + \frac{1}{4}\frac{a^2}{\delta^2}\right]$$

The equivalent circuit is now an inductance ωL in series with a resistor $\frac{1}{4}(a^2/\delta^2)\,\omega L$.

5.4. Because of the symmetry the magnetic field will be purely axial (*Figure 5.4* shows the geometry involved). At zero frequency there is no electric field and the magnetic field is, say, of strength H_0 uniform over the cross-sectional area of the coil.

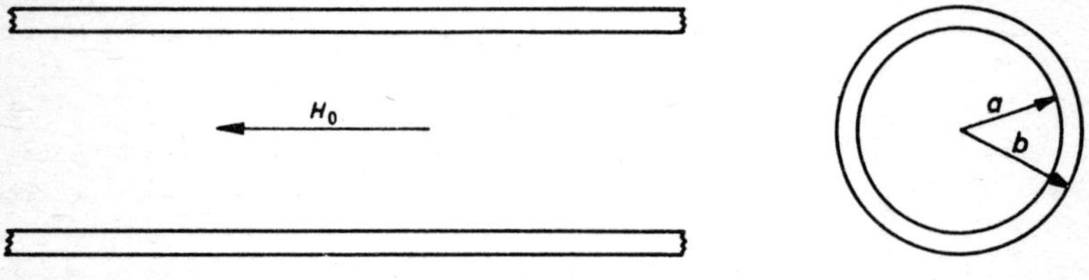

Figure 5.4.

If we now consider this magnetic field to vary harmonically in time, it will induce a circumferential electric field according to Faraday's Law; i.e.

$$\oint_{\Gamma_1} \mathbf{E}_1 \,.\, \mathrm{d}\mathbf{l} = -\frac{\mathrm{d}}{\mathrm{d}t}\int_{s_1} \mathbf{B}_0 \,.\, \mathrm{d}\mathbf{S}$$

$$\therefore \quad 2\pi r E_1 = -\, j\omega\mu_0 H_0 \pi r^2$$

$$\therefore \quad E_1 = -j\frac{\omega\mu}{2} H_0 r \qquad (r \leqslant b)$$

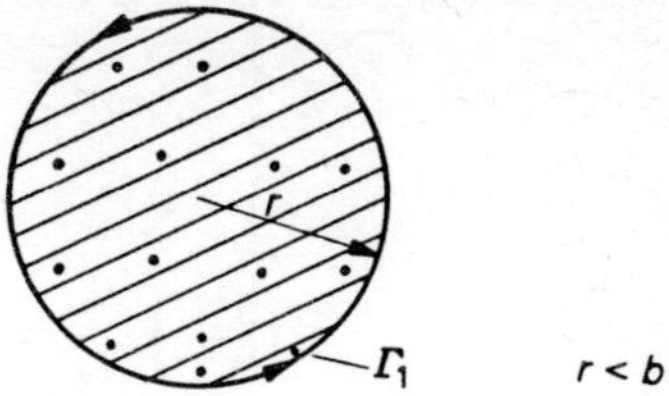

Figure 5.5.

For $a \leqslant r \leqslant b$, this electric field will cause a current to flow circumferentially inside the tube of value

$$J_1 = \sigma E_1 = -\mathrm{j}\left(\frac{\omega\mu\sigma}{2}\right) r H_0$$

$$= -\mathrm{j}\,\frac{H_0 r}{\delta^2}$$

Now this time-varying electric field (displacement current) and the conduction current in the tube will modify the magnetic field inside the tube. Since we are interested only in low frequency characteristics and because the conductivity of the tube will be rather large, we will ignore the displacement current density throughout. This assumption then implies that the field will be uniform inside the tube, i.e. $a > r$.

Let us now calculate the first order correction in the magnetic field due to J_1. Applying $\oint \mathbf{H}\,.\,\mathrm{d}\mathbf{l} = I$ to loop Γ_2.

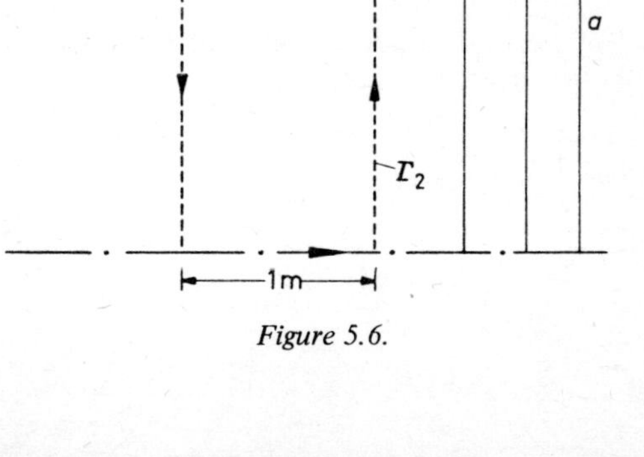

Figure 5.6.

$$\therefore \qquad H_1 = \int_a^r -J_1 \mathrm{d}r'$$

$$= \mathrm{j}\frac{H_0}{\delta^2}\int_a^r r' \,\mathrm{d}r'$$

$$= \mathrm{j}\frac{H_0}{2\delta^2}(r^2 - a^2)$$

In the above calculation we have assumed that the magnetic field on the axis of the tube is to be maintained constant at its zero frequency value hence all correction terms will be zero here.

Now this time varying magnetic field will induce an electric field, which, going back to Faraday's law, gives:

$$2\pi r E_2 = -\mathrm{j}\omega\mu_0 \int_a^r \mathrm{j}\frac{H_0}{2\delta^2}(r'^2 - a^2)\, 2\pi r' \,\mathrm{d}r'$$

$$\therefore \qquad E_2 = \frac{\omega\mu_0 H_0}{2\delta^2 r}\int_a^r \{(r')^3 - a^2 r'\}\,\mathrm{d}r'$$

$$= \frac{\omega\mu_0 H_0}{2\delta^2 r}\left[\frac{(r')^4}{4} - \frac{a^2(r')^2}{2}\right]_a^r$$

$$= \frac{\omega\mu_0 H_0}{2\delta^2 r}\left[\frac{r^4}{4} - \frac{a^2 r^2}{2} + \frac{a^4}{4}\right]$$

$$\therefore \qquad J_2 = \frac{H_0}{4\delta^4}\left[r^3 - 2a^2 r + \frac{a^4}{r}\right]$$

As previously, this second order current will produce a second order correction term in the magnetic field:

$$H_2 = \int_a^r -J_2 \,\mathrm{d}r'$$

$$= -\frac{H_0}{4\delta^4}\left[\frac{(r^4 - a^4)}{4} - a^2(r^2 - a^2) + a^4 \log_e \frac{r}{a}\right]$$

Thus at $r = b$,

$$H = H_0 + H_1 + H_2$$

$$= H_0 \left\{ 1 + \mathrm{j}\frac{(b^2 - a^2)}{2\delta^2} - \frac{1}{4\delta^4}\left[\frac{b^4}{4} + \frac{3a^4}{4} - a^2b^2 + a^4 \log_e \frac{b}{a}\right]\right\}$$

If we had taken another term into account it would have been of the order of $1/\delta^6$ and clearly we require only terms in $1/\delta^4$.

$$\therefore \qquad \left|\frac{H}{H_0}\right|^2 = 1 + \frac{1}{4\delta^4}\left[-\frac{b^4}{2} - \frac{3a^4}{2} + 2a^2b^2 - 2a^4 \log_e \frac{b}{a} + (b^2 - a^2)^2\right]$$

$$= 1 + \frac{a^4}{4\delta^4}\left[\frac{1}{2}\left(\frac{b^4}{a^4} - 1\right) - 2 \log_e \frac{b}{a}\right]$$

which is the desired result.

In the above we have used the integral laws directly for variation of approach.

5.5. Using the results from the previous problem, we have:

$$J = -\mathrm{j}\frac{H_0 r}{\delta^2} + \frac{H_0}{4\delta^4}\left[r^3 - 2a^2 r + \frac{a^4}{r}\right]$$

Now

$$P = \frac{1}{2\sigma}\int_a^b (\mathbf{J} \,.\, \mathbf{J}^*)\, 2\pi r \,\mathrm{d}r$$

Therefore considering only the first order term

$$P = \frac{\pi {H_0}^2}{\sigma\delta^4}\int_a^b r^3 \,\mathrm{d}r$$

$$\therefore \qquad P = \frac{\pi {H_0}^2}{\sigma\delta^4}\frac{(b^4 - a^4)}{4}$$

This result may be checked by application of the complex Poynting Theorem.

5.6. We shall solve this problem by the differential method. *Figure 5.4* shows the geometry with the *z*-axis of a cylindrical coordinate system being coincident with the axis of the tube.

The only field components existing will be H_z and E_ϕ. Thus Maxwell's equations in the air region inside the tube reduce to

$$\frac{dH_z}{dr} = -j\omega\epsilon E_\phi$$

$$\frac{1}{r}\frac{d}{dr}(rE_\phi) = -j\omega\mu H_z$$

and inside the wall of the tube they reduce to

$$\frac{dH_z}{dr} = -\sigma E_\phi$$

$$\frac{1}{r}\frac{d}{dr}(rE_\phi) = -j\omega\mu H_z$$

The process of solution embodies the following sequence of operations. We write both fields as power series in frequency (strictly in ω). The coefficients in these series are functions of position and if we substitute the series into Maxwell's equations, equating equal powers of frequency on both sides, we will obtain sets of differential equations linking the coefficients. There will naturally be two sets, one for the air region and one for the metal. We first solve the set in the air region and here the boundary conditions are $E_\phi(r = 0) = 0$ which is selfevident and the chosen condition that $H_z(r = 0) = H_0$, a constant for all frequencies. In effect, this latter choice means that we allow, in the solution, the magnetic field outside the tube to vary with frequency in apparent contradiction to the problem. The system is, however, a linear one so that the *ratio* of the two fields will be correct. The choice gives us an algebraic advantage as will become clear.

Having solved the equations for the air region, these solutions, evaluated at $r = a$, will give us the boundary values for the equations of the metal region, which can then be solved. Since, for this problem it is required to calculate the coefficients up to those of the fourth power in frequency the student is advised to carefully organize his algebra. In the solution here we shall omit the routine algebra for the sake of clarity.

Consider first the air region.

Let
$$E_\phi = \sum_{n=0}^{\infty} e_n\omega^n$$

and
$$H_z = \sum_{n=0}^{\infty} h_n \omega^n$$

If these series are substituted into the appropriate Maxwell's equations and the coefficients of ω^n equated on both sides there results the set of differential equations

$$\frac{dh_0}{dr} = 0$$

$$\frac{1}{r}\frac{d}{dr}(re_0) = 0$$

$$\frac{dh_n}{dr} = -j\epsilon e_{(n-1)}$$

$$\frac{1}{r}\frac{d}{dr}(re_n) = -j\mu h_{(n-1)}$$

Now for the boundary conditions. Since $E_\phi(r = 0) = 0$ for all ω then $e_n(r = 0) = 0$ for all n. Similarly, since $H_z(r = 0) = H_0$ for all ω then $h_0(r = 0) = H_0$ and $h_n(r = 0) = 0$ for all n greater than 0. We can now solve these differential equations.

Starting with the equation for e_0 it is easily shown that $e_0 = 0$ everywhere, and similarly that h_1, e_2, h_3, e_4 etc. are all zero everywhere. Thus half the coefficients vanish.

Continuing with the equation for h_0 we can easily show that $h_0 = H_0$ everywhere. Following in sequence through e_1, h_2, e_3, h_4 etc. we find

$$e_1 = -\frac{j\mu H_0}{2} r$$

$$h_2 = -\frac{\mu\epsilon H_0}{4} r^2$$

$$e_3 = \frac{j\mu^2\epsilon H_0}{16} r^3$$

$$h_4 = \frac{\mu^2\epsilon^2 H_0}{64} r^4$$

Consider now the metal region. Making the same power series substitution we obtain the following set of differential equations.

$$\frac{dh_0}{dr} = -\sigma e_0$$

$$\frac{1}{r}\frac{d}{dr}(re_0) = 0$$

$$\frac{dh_n}{dr} = -\sigma e_n$$

$$\frac{1}{r}\frac{d}{dr}(re_n) = -j\mu h_{(n-1)}$$

The boundary conditions are this time applied at $r = a$ and are obtained from the previous solutions, since E_ϕ and H_z must be continuous across the air/metal interface. The boundary conditions can be tabulated for the first five values on n as follows:

n	0	1	2	3	4
$e_n(r = a)$	0	$-j\frac{\mu a}{2}H_0$	0	$\frac{j\mu^2\epsilon a^3}{16}H_0$	0
$h_n(r = a)$	H_0	0	$-\frac{\mu\epsilon a^2}{4}H_0$	0	$\frac{\mu^2\epsilon^2 a^4}{64}H_0$

It is easily shown that e_0 is zero for all r and that $h_0 = H_0$ for all r, results that we should expect since these are effectively the d.c. terms. If we now solve for $n = 1$ and $n = 2$ we obtain

$$e_1 = -\frac{j\mu H_0}{2}r$$

$$h_1 = \frac{j\sigma\mu H_0}{4}(r^2 - a^2)$$

$$e_2 = \frac{\sigma\mu^2 H_0}{16}\left[r^3 - 2a^2 r + \frac{a^4}{r}\right]$$

$$h_2 = -\frac{\sigma^2\mu^2 H_0}{16}\left\{\frac{r^4}{4} - a^2r^2 + a^4 \ln r + \frac{3a^4}{4} - a^4 \ln a\right\} - \frac{\mu\epsilon H_0}{4}a^2$$

Now at this stage it is worth noting the difference between the two terms which appear for the first time in h_2, that is, the terms containing σ and the term containing ϵ. For a copper tube the first of these is of the order of $10^{20}a^2$ times the second which can therefore be neglected. This really amounts to ignoring all terms containing ϵ, which makes the boundary conditions for e_3 and h_4 zero. Accepting this approximation we then solve the other equations to obtain

$$e_3 = \frac{j\sigma^2\mu^3 H_0}{16}\left\{\frac{r^5}{24} - \frac{a^2r^3}{4} + \frac{a^4 r \ln r}{2} + \frac{a^4 r}{8} - \frac{a^4 r \ln a}{2} + \frac{a^6}{12r}\right\}$$

$$h_3 = -\frac{j\sigma^3\mu^3 H_0}{64}\left\{\frac{r^6}{36} - \frac{a^2r^4}{4} + a^4r^2 \ln r - \frac{a^4r^2}{4} - a^4r^2 \ln a + \frac{a^6}{3}\ln r + \frac{17}{36}a^6 - \frac{a^6}{3}\ln a\right\}$$

$$e_4 = -\frac{\sigma^3\mu^4 H_0}{64}\left\{\frac{r^7}{288} - \frac{a^2r^5}{24} + \frac{a^4r^3}{4}\ln r - \frac{a^4r^3}{8} - \frac{a^4r^3}{4}\ln a + \frac{a^6 r}{6}\ln r + \frac{11}{72}a^6 r - \frac{a^6 r}{6}\ln a + \frac{1}{96r}\right\}$$

$$h_4 = \frac{\sigma^4\mu^4 H_0}{256}\left\{\frac{r^8}{576} - \frac{a^2r^6}{36} + \frac{a^4r^4}{4}\ln r - \frac{3}{16}a^4r^4 - \frac{a^4r^4 \ln a}{4} + \frac{a^6r^2 \ln r}{3} + \frac{5}{36}a^6r^2 - \frac{a^6r^2}{3}\ln a + \frac{1}{24}\ln r + \frac{43}{576}a^8 - \frac{1}{24}\ln a\right\}$$

We now wish to evaluate H_z at $r = b$. To do this and to simplify the notation let us denote, at $r = b$,

$$h_n = j^n H_0 \alpha_n$$

Then $H_z(r = b) = H_0\,[1 + j\omega\alpha_1 - \omega^2\alpha_2 - j\omega^3\alpha_3 + \omega^4\alpha_4]$

or

$$H_0 = \frac{H_z\,(r = b)}{(1 - \alpha_2\omega^2 + \alpha_4\omega^4) + j\omega(\alpha_1 - \alpha_3\omega^2)}$$

Since we are neglecting the terms in ϵ the voltage induced in a loop of area A is just

$$V = -\mathrm{j}\omega A \mu_0 H_0$$

Now $H_z(r = b)$ is assumed to be a fixed field in practice and therefore

$$\frac{V}{H_z(r = b)} = \frac{-\mathrm{j}\omega A \mu_0}{(1 - \alpha_2 \omega^2 + \alpha_4 \omega^4) + \mathrm{j}\omega(\alpha_1 - \alpha_3 \omega^2)}$$

or, taking the modulus,

$$\left|\frac{V}{H_z(r = b)}\right|^2 = \frac{\omega^2 {\mu_0}^2 A^2}{1 + \omega^2({\alpha_1}^2 - 2\alpha_2) + \omega^4({\alpha_2}^2 + 2\alpha_4 - 2\alpha_1\alpha_3)}$$

neglecting terms of order greater than ω^4.

This expression has a maximum when

$$\omega^4 = [{\alpha_2}^2 + 2\alpha_4 - 2\alpha_1\alpha_3]^{-1}$$

and the problem is effectively solved. The values of α_n are (with $x = (b/a)$)

$$\alpha_1 = \frac{\sigma\mu a^2}{4}(x^2 - 1)$$

$$\alpha_2 = \frac{\sigma^2\mu^2 a^4}{16}\left(\frac{1}{4}x^4 - x^2 + \frac{3}{4} + \ln x\right)$$

$$\alpha_3 = \frac{\sigma^3\mu^3 a^6}{64}\left(\frac{1}{36}x^6 \quad \frac{1}{4}x^4 + x^2 \ln x - \frac{1}{4}x^2 + \frac{1}{3}\ln x + \frac{17}{36}\right)$$

$$\alpha_4 = \frac{\sigma^4\mu^4 a^8}{256}\left(\frac{1}{576}x^8 - \frac{1}{36}x^6 - \frac{3}{16}x^4 + \frac{5}{36}x^2 + \frac{1}{4}x^4 \ln x + \frac{1}{3}x^2 \ln x + \frac{1}{24}\ln x + \frac{43}{576}\right)$$

5.7. In this problem we are told to take a flat plate model of the system. The geometry of this is shown in *Figure 5.7* in which the cartesian z-axis replaces the radial coordinate of the cylindrical system. In the tubular structure the components of **E** and **H** were tangential to the metal surface so

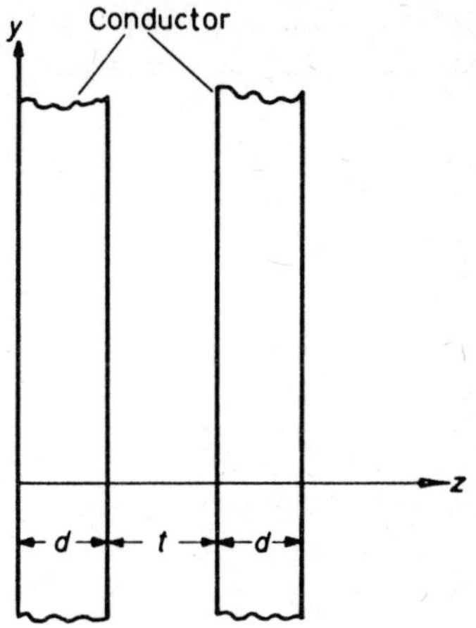

Figure 5.7.

we shall assume that these components become E_y and H_x in the model. We compare the plane $z = 0$ with the inner surface of the inner tube and the plane $z = (2d + t)$ with the outer surface of the outer tube.

As in the previous problem we expand the fields in series of powers of ω, substitute into the appropriate Maxwell's equation for the region and derive a set of differential equations in the coefficients. By these means we obtain the following sets:

(*i*) *Air region*

$$\frac{de_0}{dz} = 0; \qquad \frac{dh_0}{dz} = 0$$

$$\frac{de_n}{dz} = j\mu h_{(n-1)}; \qquad \frac{dh_n}{dz} = j\epsilon e_{(n-1)}$$

(*ii*) *Conducting region*

$$\frac{de_0}{dz} = 0; \qquad \frac{dh_0}{dz} = \sigma e_0$$

$$\frac{de_n}{dz} = j\mu h_{(n-1)}; \qquad \frac{dh_n}{dz} = \sigma e_n$$

We now start to solve these from the 'inside' to the 'outside', in this case from $z = 0$ to $z = (2d + t)$. It is clear that everywhere $e_0 \equiv 0$ and $h_0 = H_0$ (say) since these are d.c. conditions.

First conducting region–The boundary conditions are $e_n(0) = 0$; $h_n(0) = 0$ for all n of interest ($\geqslant 1$).

The differential equations are trivial and we obtain the values of the coefficients (to second order) at $z = d$ as

$$h_0(d) = H_0 \qquad e_0(d) = 0$$

$$h_1(d) = \mathrm{j}\sigma\mu H_0 \frac{d^2}{2} \qquad e_1(d) = \mathrm{j}\mu H_0 d$$

$$h_2(d) = -\sigma^2\mu^2 H_0 \frac{d^4}{24} \qquad e_2(d) = -\sigma\mu^2 H_0 \frac{d^3}{6}$$

and these are the boundary conditions for the air region.

Air region–Again the equations are trivial leading to the values of the coefficients at $z = (d + t)$ as

$$h_0(d + t) = H_0 \qquad e_0(d + t) = 0$$

$$h_1(d + t) = \mathrm{j}\sigma\mu H_0 \frac{d^2}{2} \qquad e_1(d + t) = \mathrm{j}\mu H_0(d + t)$$

$$h_2(d + t) = -\sigma^2\mu^2 H_0 \frac{d^4}{24} - \mu\epsilon H_0 t\left(d + \frac{t}{2}\right) \qquad e_2(d + t) = -\sigma\mu^2 H_0 \frac{d^2}{2}\left(t + \frac{d}{3}\right)$$

which are the boundary conditions for the second conducting region.

Second conducting region–The solution of these equations leads to the values of the magnetic field coefficients (which are the only ones now of interest) at $z = (2d + t)$ as

$$h_0(2d + t) = H_0$$

$$h_1(2d + t) = \mathrm{j}\sigma\mu H_0 d(2d + t)$$

$$h_2(2d + t) = -\sigma^2\mu^2 H_0 \frac{2}{3} d^3(d + t) - \mu\epsilon H_0 t\left(d + \frac{t}{2}\right) \approx -\sigma^2\mu^2 H_0 \frac{2}{3} d^3(d + t)$$

Thus the total magnetic field at $(2d + t)$ compared with H_0 is

$$\frac{H_x(2d + t)}{H_0} = 1 + \mathrm{j}\omega\sigma\mu d(2d + t) - \omega^2\sigma^2\mu^2 \frac{2}{3} d^3(d + t)$$

and taking the modulus to second order is

$$\left|\frac{H_x(2d+t)}{H_0}\right|^2 = 1 + \omega^2\sigma^2\mu^2\frac{d^2}{3}(8d^2 + 8dt + 3t^2)$$

and this can be regarded as a screening factor. The corresponding factor for the single conductor of double thickness can easily be obtained from this by putting $t = 0$, since extra air space either inside or outside has little effect, when

$$\left|\frac{H_x(2d)}{H_0}\right|^2 = 1 + \omega^2\sigma^2\mu^2\frac{8d^4}{3}$$

If we define the skin depth δ then the ratio of these two screening factors can be written as

$$\frac{\text{(Screening factor, two plates)}}{\text{(Screening factor, one plate)}} = \frac{1 + \dfrac{32}{3}\left(\dfrac{d}{\delta}\right)^4\left(1 + \dfrac{t}{d} + \dfrac{3t^2}{8d^2}\right)}{1 + \dfrac{32}{3}\left(\dfrac{d}{\delta}\right)^4}$$

and we find that in fact the double plate system is more effective than is the single plate of double thickness. It is useful to consider a numerical example. Suppose that the plates are one skin depth thick and four skin depths apart. The above ratio is easily shown to be slightly over 10 in magnitude. ($d = \delta$, $t = 4\delta$).

5.8. This problem has the same geometry as Problem **5.1** but we shall solve it using the differential equation approach.

In this case the field components are E_z and H_ϕ and Maxwell's equations reduce to

$$\frac{\mathrm{d}E_z}{\mathrm{d}r} = \mathrm{j}\omega\mu H_\phi$$

$$\frac{1}{r}\frac{\mathrm{d}}{\mathrm{d}r}(rH_\phi) = \sigma E_z$$

Substituting series expansions in ω for the electric and magnetic fields as in the previous two problems we obtain a set of coefficient differential equations

$$\frac{de_0}{dr} = 0$$

$$\frac{de_n}{dr} = j\mu h_{(n-1)}$$

$$\frac{1}{r}\frac{d}{dr}(rh_n) = \sigma e_n$$

As boundary conditions we choose $H_\phi(r = 0) = 0$ and $E_z(r = 0) = E_0$ for all frequencies. This means that the coefficients must satisfy

$$e_0(r = 0) = E_0$$

$$e_n(r = 0) = 0$$

$$h_n\ (r = 0) = 0 \text{ for all } n$$

If these differential equations are now solved up to and including $n = 2$, we obtain

$$e_0 = E_0 \qquad h_0 = \frac{\sigma E_0}{2} r$$

$$e_1 = j\frac{\mu\sigma E_0}{4} r^2 \qquad h_1 = \frac{j\mu\sigma^2 E_0}{16} r^3$$

$$e_2 = -\frac{\mu^2\sigma^2 E_0}{64} r^4 \qquad h_2 = -\frac{\mu^2\sigma^3 E_0}{384} r^5$$

There now remains the question of calculating the internal self-inductance. This we shall do using the two forms of expression for the time-average stored energy in an inductor. Thus

$$\tfrac{1}{4} L\,|I|^2 = \tfrac{1}{4}\mu \int_{\text{unit length}} |H|^2 dV$$

We therefore need the full expressions for E and H. Thus

$$E_z = E_0\left\{1 + \mathrm{j}\frac{\omega\mu\sigma}{4}r^2 - \frac{\omega^2\mu^2\sigma^2}{64}r^4\right\}$$

and

$$H_\phi = \frac{\sigma E_0}{2}\left\{r + \mathrm{j}\frac{\omega\mu\sigma}{8}r^3 - \frac{\omega^2\mu^2\sigma^2}{192}r^5\right\}$$

From the first of these the current density J_z is just

$$J_z = \sigma E_0\left\{1 + \mathrm{j}\frac{\omega\mu\sigma}{4}r^2 - \frac{\omega^2\mu^2\sigma^2}{64}r^4\right\}$$

and hence the total current I is given by

$$I = \int_{r=0}^{a} J_z 2\pi r\,\mathrm{d}r = \pi a^2\sigma E_0\left\{1 + \mathrm{j}\frac{\omega\mu\sigma}{8}a^2 - \frac{\omega^2\mu^2\sigma^2}{192}a^4\right\}$$

whence

$$|I|^2 = [\pi a^2\sigma E_0]^2\left[1 + \frac{\omega^2\mu^2\sigma^2}{192}a^4\right] \text{ to order } \omega^2$$

From the magnetic field

$$\int |H|^2\,\mathrm{d}v = \left(\frac{\sigma E_0}{2}\right)^2\int_0^a\left[r^2 + \frac{\omega^2\mu^2\sigma^2}{192}r^6\right]2\pi r\,\mathrm{d}r = \frac{\pi a^4\sigma^2 E_0{}^2}{8}\left\{1 + \frac{\omega^2\mu^2\sigma^2}{384}a^4\right\}$$

and therefore the internal self-inductance per unit length is given by

$$L = \frac{\mu}{8\pi}\frac{\left[1 + \dfrac{\omega^2\mu^2\sigma^2}{384}a^4\right]}{\left[1 + \dfrac{\omega^2\mu^2\sigma^2}{192}a^4\right]} = \frac{\mu}{8\pi}\left[1 - \frac{\omega^2\mu^2\sigma^2}{384}a^4\right]$$

or, inserting the skin depth parameter,

$$\delta = \sqrt{\frac{2}{\omega\mu\sigma}},$$

$$L = \frac{\mu}{8\pi}\left[1 - \frac{1}{96}\left(\frac{a}{\delta}\right)^4\right]$$

CHAPTER 6

6.1. Making the same assumptions as in Problem 5.2 Maxwell's curl relations reduce to

$$\frac{\partial H_y}{\partial x} = \sigma E_z \qquad \ldots .(1)$$

$$\frac{\partial E_z}{\partial x} = j\omega\mu H_y \qquad \ldots .(2)$$

Eliminating H_y from Equations (1) and (2):

$$\frac{\partial^2 E_z}{\partial x^2} = j\omega\mu\sigma E_z$$

the solution of which we can write down immediately by inspection in the form:

$$E_z = A\, e^{(1+i)(x/\delta)} + B\, e^{(1+j)(x/\delta)} \qquad \ldots .(3)$$

where δ is $\dfrac{1}{\sqrt{\pi f \mu \sigma}}$, the skin depth factor.

The boundary conditions imposed on the field vectors require that the tangential component of H at the copper–iron interface be continuous and since it is zero in the iron, it must be zero in the copper, i.e.

$$H_y = \frac{1}{j\omega\mu} \cdot \frac{\partial E_z}{\partial x} = 0 \text{ at } x = 0$$

Thus from Equation (3),

$$\frac{(1+j)}{\delta}[A - B] = 0$$

$$\therefore \qquad A = B$$

Therefore the current density at any point in the conductor is given by

$$J_z = \sigma E_z = \sigma A\,[e^{(1+j)(x/\delta)} + e^{-(1+j)(x/\delta)}]$$

We can get the total current flowing by integrating J_z over the cross section, i.e.

$$I = \sigma A W \int_0^h (e^{(1+j)(x/\delta)} + e^{-(1+j)(x/\delta)})\,dx$$

$$= \frac{\sigma A W\delta}{(1+j)}[e^{(1+j)s} - e^{-(1+j)s}] \text{ where } s = \frac{h}{\delta}$$

$$\therefore \qquad I = \frac{2\sigma A W\delta}{(1+j)} \sinh(s + js)$$

If this total current flowed through the d.c. resistance of the conductor, the loss would be:

$$P_{d.c.} = \frac{1}{2}\left[\frac{4\sigma^2 A^2 W^2 \delta^2}{2}(\sinh^2 s \,.\, \cos^2 s + \cosh^2 s \sin^2 s)\right]\frac{1}{\sigma W h}$$

which on rearrangement of the terms gives

$$P_{d.c.} = \frac{\sigma W A^2 \delta^2}{2h}[\cosh 2s - \cos 2s] \text{ W/m} \qquad \ldots . (4)$$

We can now calculate the actual loss in the conductor and by comparing this with Equation (4) derive an expression for the a.c. resistance. Power loss per unit volume $= \frac{1}{2}\sigma \mathbf{E} \,.\, \mathbf{E}^*$

$$= \tfrac{1}{2}\sigma A^2 (e^{x/\delta} \,.\, e^{j(x/\delta)} + e^{-(x/\delta)} \,.\, e^{-j(x/\delta)})(e^{x/\delta} \,.\, e^{-j(x/\delta)} + e^{-(x/\delta)} \,.\, e^{j(x/\delta)})$$

$$= \tfrac{1}{2}\sigma A^2 (e^{2x/\delta} + e^{-j(2x/\delta)} + e^{j(2x/\delta)} + e^{-(2x/\delta)})$$

$$= \sigma A^2 \left(\cosh \frac{2x}{\delta} + \cos \frac{2x}{\delta}\right)$$

$$\therefore P_{a.c.} = \sigma A^2 W \int_0^h \left(\cosh \frac{2x}{\delta} + \cos \frac{2x}{\delta}\right) dx$$

$$\therefore \qquad P_{\text{a.c.}} = \frac{\sigma A^2 W\delta}{2}(\sinh 2s + \sin 2s) \text{ W/m} \qquad \ldots (5)$$

Hence

$$\frac{P_{\text{a.c.}}}{P_{\text{d.c.}}} = \frac{R_{\text{a.c.}}}{R_{\text{d.c.}}} = s\left(\frac{\sinh 2s + \sin 2s}{\cosh 2s - \cos 2s}\right) \qquad \ldots (6)$$

If we now expand sinh, sin, cosh and cos as power series and retain only the first few terms, we should recover our approximate result we obtained in the previous chapter.

$$\sinh(2s) = 2s + \frac{8s^3}{6} + \frac{(2s)^5}{5!}$$

$$\sin(2s) = 2s - \frac{8s^3}{6} + \frac{(2s)^5}{5!}$$

$$\therefore \qquad [\sinh(2s) + \sin(2s)]\,s = 4s^2 + \frac{8}{15}s^6$$

$$= 4s^2\left(1 + \frac{2}{15}s^4\right)$$

$$\cosh(2s) = 1 + \frac{(2s)^2}{2} + \frac{(2s)^4}{24} + \frac{(2s)^6}{6!}$$

$$\cos(2s) = 1 - \frac{(2s)^2}{2} + \frac{(2s)^4}{24} - \frac{(2s)^6}{6!}$$

$$\therefore \qquad \cosh(2s) - \cos(2s) = (2s)^2 + 2\frac{(2s)^6}{6!}$$

$$= 4s^2\left(1 + \frac{2}{45}s^4\right)$$

$$s\left(\frac{\sinh(2s) + \sin(2s)}{\cosh(2s) - \cos(2s)}\right) \doteqdot \left(1 + \frac{6}{45}s^4\right)\left(1 + \frac{2}{45}s^4\right)^{-1}$$

$$= \left(1 + \frac{4}{45}s^4\right)$$

6.2. Making the same physical assumptions we did in the solution to Problem **5.3**, we now wish to determine the circumferential electric field at the surface of the rod, without recourse to quasi-static methods, in terms of the longitudinal magnetic field at the rod surface.

The wave equation for **H** within the rod is

$$\nabla^2\mathbf{H} = \mathrm{j}\omega\mu\sigma\mathbf{H}$$

Expanding in terms of cylindrical polar coordinates gives

$$\frac{1}{r}\frac{\partial}{\partial r}\left(r\frac{\partial H_z}{\partial r}\right) = \mathrm{j}\omega\mu\sigma H_z$$

or

$$r^2\frac{\mathrm{d}^2H_z}{\mathrm{d}r^2} + r\frac{\mathrm{d}H_z}{\mathrm{d}r} - \mathrm{j}\omega\mu\sigma H_z r^2 = 0$$

the solution of which can be put in terms of zero order Bessel Functions. The Bessel Function of the second kind is not an allowable solution in this problem because the field on the axis must remain finite.

$$\therefore \qquad H_z = AJ_0(kr) \qquad [k \equiv (-\mathrm{j}\omega\mu\sigma)^{1/2}]$$

At $r = a$, $H_z = H_0$ say

$$\therefore \qquad A = \frac{H_0}{J_0(ka)}$$

$$\therefore \qquad H_z = H_0\frac{J_0(kr)}{J_0(ka)}$$

Using

$$\nabla \times \mathbf{H} = \sigma\mathbf{E}$$

we get

$$-\frac{\partial H_z}{\partial r} = \sigma E_\phi$$

$$\therefore \qquad E_\phi = -\frac{H_0}{\sigma}k\frac{J_0'(kr)}{J_0(ka)} = H_0\frac{k}{\sigma}\frac{J_1(kr)}{J_0(ka)}$$

Integrating $(-E_\phi)$ at $r = a$ along the length of the wire in the coil gives us the applied voltage, V, i.e.

$$V = -H_0 \frac{k}{\sigma} \frac{J_1(ka)}{J_0(ka)} \cdot 2\pi a N$$

Now
$$H_0 = \frac{NI}{l}$$

$$\therefore \quad \underline{Z = \frac{V}{I} = -\frac{k}{\sigma} \frac{J_1(ka)}{J_0(ka)} \frac{2\pi a N^2}{l}}$$

For small values of the argument (ka):

$$J_1(ka) \approx \frac{ka}{2}, \qquad J_0(ka) \approx 1$$

$$\therefore \quad Z \doteqdot -\frac{k}{\sigma} \frac{ka}{2} \cdot 2\pi a \frac{N^2}{l}$$

$$= j\omega\mu \cdot \frac{\pi a^2 N^2}{l}$$

$$\therefore \quad \underline{Z \doteqdot j\omega L} \qquad \left(L \equiv \frac{\mu \pi a^2 N^2}{l}\right)$$

To the second order of approximation:

$$J_1(ka) \doteqdot \frac{ka}{2}\left(1 - \frac{(ka)^2}{8}\right)$$

$$J_0(ka) \doteqdot 1 - \frac{(ka)^2}{4}$$

$$\therefore \quad Z = -\frac{k}{\sigma} \cdot \frac{ka}{2} \cdot \frac{\left(1 - \frac{(ka)^2}{8}\right)}{\left(1 - \frac{(ka)^2}{4}\right)} \cdot 2\pi a \frac{N^2}{l}$$

$$\doteqdot j\omega L \left(1 - \frac{(ka)^2}{8}\right)\left(1 + \frac{(ka)^2}{4}\right)$$

$$= j\omega L \left[1 - j\frac{\omega\mu\sigma a^2}{8}\right]$$

$$\therefore \quad \underline{Z \doteqdot \omega L \left[j + \frac{1}{4}\frac{a^2}{\delta^2}\right]}$$

We have now recovered our approximate expression for the impedance of the coil.

6.3. If the skin depth is very much less than the radius of the conductor, then we can to a first approximation ignore the surface curvature (see *Figure 6.2*). We then solve the simpler problem of the current flow in a semi-infinite conductor.

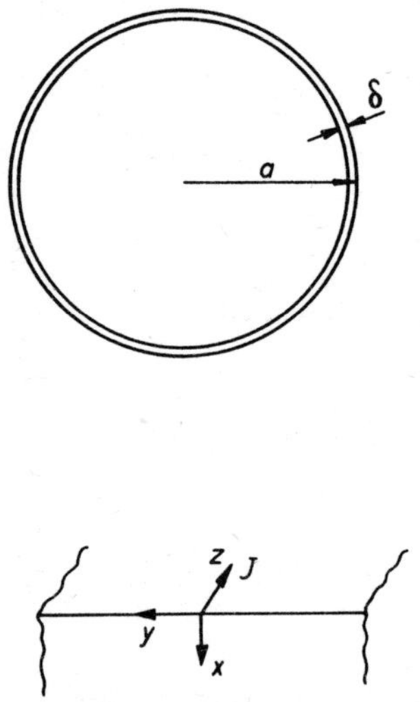

Figure 6.2.

With the chosen set of coordinates, the governing equation is

$$\frac{d^2 J_z}{dx^2} = j\omega\mu\sigma J_z$$

where we are assuming time variations of the form $e^{j\omega t}$. Hence

$$J_z = A\, e^{-(1+j)(x/\delta)} + B\, e^{(1+j)(x/\delta)} \qquad \left(\delta \equiv \frac{1}{\sqrt{\pi f \mu\sigma}}\right)$$

The first term represents a wave travelling along the positive x axis whilst the second term represents a wave travelling along the negative x axis. For a semi-infinite solid, there can be no reflected wave

$$\therefore \qquad B = 0$$

Hence

$$J_z = A\, e^{-(1+j)(x/\delta)}$$

Now the Joule heat loss is simply $(1/2\sigma)\,|J|^2$ per unit volume; therefore, the energy loss per unit area of surface

$$= \frac{A^2}{2\sigma}\int_0^\infty e^{-2x/\delta}\, dx$$

$$= \frac{A^2\delta}{4\sigma}$$

Hence at high frequencies, when the curvature of the conductor can be neglected,

$$\text{total power loss per metre, } P_{\text{a.c.}} = \frac{A^2\delta}{4\sigma} \cdot 2\pi a \qquad \ldots.(1)$$

We now need to know the total current flowing in the wire so that we can calculate what loss would occur if this current flowed through the d.c. resistance of the wire.

From the expression for the current density above, we can find the current per unit width of conductor, i.e.

$$\int_0^\infty J\,dx$$

$$= \int_0^\infty A\,e^{-(1+j)(x/\delta)}\,dx$$

$$= \frac{A\delta}{(1+j)}$$

Therefore, total current in wire, $I = \dfrac{A\delta}{(1+j)}\,2\pi a$ ....(2)

Hence d.c. power loss,

$$P_{d.c.} = \tfrac{1}{2}\,|I|^2\,\frac{1}{\sigma\pi a^2} \text{ per metre length}$$

$$= \frac{1}{2}\cdot\frac{A^2\delta^2}{2}\cdot\frac{4\pi^2 a^2}{\sigma\pi a^2}$$

$$= \frac{A^2\delta^2\pi}{\sigma}$$

$$\frac{P_{a.c.}}{P_{d.c.}} = \frac{R_{a.c.}}{R_{d.c.}} = \left(\frac{A^2\delta}{4\sigma}\cdot 2\pi a\right)\frac{\sigma}{A^2\delta^2\pi}$$

$$\therefore \quad \underline{\frac{R_{a.c.}}{R_{d.c.}} = \frac{a}{2\delta}}$$

Now the high frequency losses are directly proportional to $R_{a.c.}$ and we see that this is proportional to $\sigma^{-1/2}$. Thus if we reduce σ^{-1} to 1/100th of its uncooled value, the losses will be reduced to 1/10th of their uncooled value.

6.4. The equation governing the magnetic field within the sheet is given by

$$\frac{\partial^2 H}{\partial y^2} = \frac{\mu_0 \mu_r}{\rho} \frac{\partial H}{\partial t}$$

which for an exponential time dependence, $e^{j\omega t}$, reduces to

$$\frac{d^2 H}{dy^2} = \frac{j\omega\mu_0\mu_r}{\rho} H$$

the solution to which is

$$H = A\, e^{(1+j)(y/\delta)} + B\, e^{-(1+j)(y/\delta)} \qquad \left(\delta \equiv \sqrt{\frac{\rho}{\pi f \mu_0 \mu_r}}\right)$$

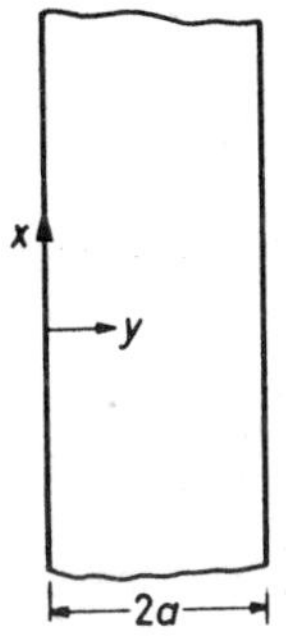

Figure 6.3.

Let H_s be the surface value of H, i.e. at $y = 0$, $y = 2a$ (see *Figure 6.3*).

$$H_s = A + B = A\, e^{(1+j)2s} + B\, e^{-(1+j)2s} \qquad \left(s \equiv \frac{a}{\delta}\right)$$

$$\therefore \quad A\, e^{(1+j)s}\left[e^{(1+j)s} - e^{-(1+j)s}\right] = B\, e^{-(1+j)s}\left[e^{(1+j)s} - e^{-(1+j)s}\right]$$

$$\therefore \quad B = A\, e^{(1+j)2s}$$

$$\therefore \quad H_s = A\left[1 + e^{(1+j)2s}\right]$$

$$\therefore \qquad A = \frac{H_s}{1 + e^{(1+j)2s}}$$

$$H = \frac{H_s}{(1 + e^{(1+j)2s})}\left[e^{(1+j)(y/\delta)} + e^{(1+j)2s}\, e^{-(1+j)(y/\delta)}\right]$$

$$H_0 = \frac{H_s}{(1 + e^{(1+j)2s})}\left[e^{(1+j)s} + e^{(1+j)s}\right]$$

$$= \frac{H_s}{\cosh (s + js)}$$

$$\therefore \qquad \frac{H_s}{H_0} = \cosh (s + js) = \cosh .s \cos .s + j \sinh .s \sin .s$$

$$\therefore \qquad \left|\frac{H_s}{H_0}\right|^2 = \cosh^2 s \cos^2 s + \sinh^2 s \sin^2 s = ½ [\cosh 2s + \cos 2s]$$

6.5. The equation governing the current density in the sheet is

$$\frac{d^2J}{dy^2} - \frac{j}{d^2} J = 0$$

the solution of which can be put in the form

$$J_z = A \cosh \frac{(1 + j)y}{d\sqrt{2}} + B \sinh \frac{(1 + j)y}{d\sqrt{2}} \qquad \ldots .(1)$$

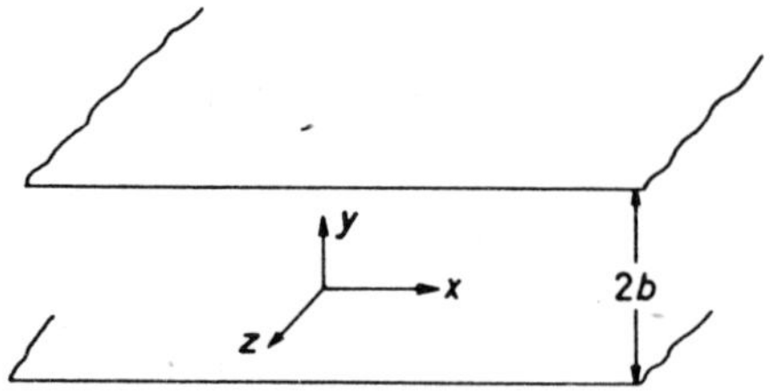

Figure 6.4.

where for convenience we have let J_z only have a z component. At the centre of the strip (i.e. $y = 0$),

$$J_z = J_0$$

$$\therefore \qquad A = J_0 \text{ (r.m.s.)}$$

If we now assume that both faces of the strip are being identically excited then

$$J_z(b) = J_z(-b)$$

and the current distribution will be symmetrical about the centre plane. This implies that $B = 0$ since sinh is an odd function. Hence

$$J_z = J_0 \cosh \frac{(1 + \mathrm{j})y}{d\sqrt{2}}$$

$$\therefore \qquad E_z = \rho\, J_0 \cosh \frac{(1 + \mathrm{j})y}{d\sqrt{2}} \qquad \ldots (2)$$

where ρ is the resistivity of the strip.

Using

$$\nabla \times \mathbf{E} = -\mathrm{j}\omega\mu\mathbf{H}$$

i.c.

$$\frac{\partial E_z}{\partial y} = -\mathrm{j}\omega\mu H_x$$

we obtain

$$\frac{\partial E_z}{\partial y} = \rho\, J_0 \frac{(1 + \mathrm{j})}{d\sqrt{2}} \sinh \frac{(1 + \mathrm{j})y}{d\sqrt{2}} = -\mathrm{j}\omega\mu H_x$$

$$\therefore \qquad H_x = \mathrm{j}\, J_0\, d \frac{(1 + \mathrm{j})}{\sqrt{2}} \sinh \frac{(1 + \mathrm{j})y}{d\sqrt{2}}$$

$$\therefore \qquad H_x = -J_0\, d \frac{(1 - \mathrm{j})}{\sqrt{2}} \sinh \frac{(1 + \mathrm{j})y}{d\sqrt{2}} \qquad \ldots (3)$$

Putting $y = b$ and using the substitution

$$\theta = \frac{b}{\mathrm{d}\sqrt{2}}$$

we have the fields at the conductor surface, i.e.

$$E_z = \rho J_0 \cosh(1 + \mathrm{j})\theta = \rho J_0 \{\cosh\theta\cos\theta + \mathrm{j}\sinh\theta\sin\theta\}$$

$$H_x^* = -J_0 \mathrm{d}\frac{(1+\mathrm{j})}{\sqrt{2}}\sinh(1-\mathrm{j})\theta$$

$$= -\frac{J_0\mathrm{d}}{\sqrt{2}}(1+\mathrm{j})\{\sinh\theta\cos\theta - \mathrm{j}\cosh\theta\sin\theta\}$$

Now the loss per unit area of each face is simply the real part of the complex Poynting vector. Therefore

$$\text{total loss} = -2\,\mathrm{Re}\,[E_z H_x^*]$$

$$= \rho J_0^2 \mathrm{d}\sqrt{2}\{\mathrm{Re}[(1+\mathrm{j})(\sinh\theta\cosh\theta\cos^2\theta + \sinh\theta\cosh\theta\sin^2\theta$$

$$+ \mathrm{j}\sin\theta\cos\theta\sinh^2\theta - \mathrm{j}\sin\theta\cos\theta\cosh^2\theta)]\}$$

$$= \rho J_0^2 \mathrm{d}\sqrt{2}\,\mathrm{Re}[(1+\mathrm{j})(\sinh\theta\cosh\theta - \mathrm{j}\sin\theta\cos\theta)]$$

$$= \rho J_0^2 \mathrm{d}\sqrt{2}\,[\sinh\theta\cosh\theta + \sin\theta\cos\theta].$$

6.6. (*a*) By a 'good conductor' we mean one in which the displacement current density is small compared to the conduction current density; hence ignoring the displacement current term, Maxwell's equations can be reduced to

$$\nabla \times \mathbf{E} = -\mathrm{j}\omega\mu\mathbf{H}$$

$$\nabla \times \mathbf{H} = \sigma\mathbf{E}$$

$$\nabla \cdot \mathbf{E} = 0$$

$$\nabla \cdot \mathbf{H} = 0$$

where we have further assumed a time dependence of the form $e^{\mathrm{j}\omega t}$.

The argument now proceeds along lines similar to those used in Problem 4.3. If **n** is the unit vector in the direction of propagation of the plane wave, then ∇ becomes $\mathbf{n}\dfrac{\partial}{\partial z}$. Hence

$$\nabla \cdot \mathbf{E} = \frac{\partial}{\partial z}(\mathbf{n} \cdot \mathbf{E}) = 0$$

and

$$\nabla \cdot \mathbf{H} = \frac{\partial}{\partial z}(\mathbf{n} \cdot \mathbf{H}) = 0$$

signify that both **E** and **H** are transverse to the direction of propagation. Now

$$\nabla \times \mathbf{E} = \mathbf{n} \times \frac{\partial \mathbf{E}}{\partial z} = -\mathrm{j}\omega\mu\mathbf{H}$$

$$\therefore \qquad \mathbf{n} \times \left(\mathbf{n} \times \frac{\partial^2 \mathbf{E}}{\partial z^2}\right) = -\frac{\partial^2 \mathbf{E}}{\partial z^2} = -\mathrm{j}\omega\mu\sigma\mathbf{E}$$

$$\therefore \qquad \mathbf{E} = \mathscr{E}^+ \mathrm{e}^{-\gamma z} + \mathscr{E}^- \mathrm{e}^{\gamma z}$$

where the propagation constant $\gamma = \alpha + \mathrm{j}\beta = \sqrt{\mathrm{j}\omega\mu\sigma}$

Hence

$$\text{the phase constant, } \beta = \mathrm{Im}[\gamma] = \mathrm{Im}\left[\frac{(1+\mathrm{j})}{\sqrt{2}}\sqrt{\omega\mu\sigma}\right] = \sqrt{\frac{\omega\mu\sigma}{2}}$$

$$\text{and attenuation constant, } \alpha = \mathrm{Re}[\gamma] = \sqrt{\frac{\omega\mu\sigma}{2}} = \beta$$

Now the skin depth factor is that distance which either a positive or negative going wave must travel in order to decrease in amplitude by a factor e^{-1}.

$$\therefore \qquad \delta = \frac{1}{\sqrt{\pi f \mu\sigma}}$$

The phase velocity is given by

$$v_p = \frac{\omega}{\beta} = \sqrt{\frac{2\omega}{\mu\sigma}}$$

and the group velocity by

$$v_g = \frac{d\omega}{d\beta} = \sqrt{\frac{8\omega}{\sigma\mu}}$$

Now $$\frac{\partial \mathbf{E}}{\partial z} = -\gamma\,[\mathscr{E}^{+}e^{-\gamma z} - \mathscr{E}^{-}e^{\gamma z}]$$

$$\therefore \quad \mathbf{H} = \frac{1}{-j\omega\mu}\left(\mathbf{n} \times \frac{\partial \mathbf{E}}{\partial z}\right)$$

$$= \frac{1}{\left(j\dfrac{\omega\mu}{\gamma}\right)}\,\mathbf{n} \times [\mathscr{E}^{+}e^{-\gamma z} - \mathscr{E}^{-}e^{\gamma z}]$$

$$\therefore \quad \frac{\mathscr{E}^{+}}{\mathscr{H}^{+}} = \frac{\mathscr{E}^{-}}{\mathscr{H}^{-}} = j\frac{\omega\mu}{\gamma} = \eta \text{ (intrinsic impedance)}$$

$$\therefore \quad \eta = \sqrt{j\frac{\omega\mu}{\sigma}} = (1+j)\sqrt{\frac{\pi f\mu}{\sigma}}$$

(*b*) $$\beta = \alpha = \frac{1}{\delta} = \sqrt{\frac{\omega\mu\sigma}{2}}$$

$$\therefore \quad \delta = \frac{1}{\sqrt{\pi f\mu\sigma}} = \frac{1}{[\pi\, 50 \times 2 \times 10^6 \times 4\pi\, 10^{-7} \times 2500]^{1/2}} \risingdotseq 0{\cdot}001 \text{ m}$$

i.e. 1 mm

$\therefore\ \beta = \alpha = 1000$

$$v_p = \frac{\omega}{\beta} = \frac{314}{1000} = 0{\cdot}314 \text{ m/s}$$

$$v_g = \sqrt{\frac{8\omega}{\mu\sigma}} = \sqrt{0{\cdot}4} = 0{\cdot}633 \text{ m/s}$$

$$\eta = (1 + \text{j})\sqrt{\frac{\pi\ 50 \times 2500 \times 4\pi\ 10^{-7}}{2 \times 10^6}} = (1 + \text{j})(5 \times 10^{-4})$$

We note from the intrinsic impedance that the magnetic field lags the electric field by 45° and that the energy in the wave resides almost entirely in the magnetic field.

Another interesting point worth noting is that the skin depth is quite small compared with the dimensions of say, a transformer that would use such a material as this. This is the reason why a transformer is laminated and not, as is often erroneously quoted in some text books, because it reduces the eddy current losses. If it were not laminated the magnetic field would never penetrate into the mass of the steel and so this must be the prime reason for laminating. True enough, making the material even thinner, once one has laminated it, will admittedly reduce the losses—an important but secondary effect.

6.7. *Region (2)*—Because region (2) is semi-infinite we will have only one wave propagating and this will be along the positive x axis, the components of which, E_2 and H_2 are everywhere related by the intrinsic impedance of the medium, i.e.

$$E_2 = E_0\, \text{e}^{-\gamma_2 x} = E_0\, \text{e}^{-(1+\text{j})\sqrt{\pi f \mu_2 \sigma_2}\, x}$$

$$H_2 = \frac{E_0}{\eta_2}\text{e}^{-\gamma_2 x} = \frac{E_0}{\eta_2}\, \text{e}^{-(1+\text{j})\sqrt{\pi f \mu_2 \sigma_2}\, x}$$

$$\gamma_2^2 = \text{j}\omega\mu_2\sigma_2$$

$$\eta_2 = \sqrt{\frac{\text{j}\omega\mu_2}{\sigma_2}} \quad \text{(see Problem } \mathbf{6.6}\text{)}$$

As usual we have omitted the exponential time dependence.

Hence $$J_2 = J_0\, e^{-\sqrt{\pi f \mu_2 \sigma_2}\,x}\, e^{-j\sqrt{\pi f \mu_2 \sigma_2}\,x}$$

Region (1)–Let us now apply the *ABCD* matrix developed in Problem **4.8** which relates the total fields between two reference planes distance 'l' apart, i.e.

$$\begin{bmatrix} E_s \\ H_s \end{bmatrix} = \begin{bmatrix} \cosh \gamma_1 l & \eta_1 \,.\, \sinh \gamma_1 l \\ \dfrac{1}{\eta_1} \,.\, \sinh \gamma_1 l & \cosh \gamma_1 l \end{bmatrix} \begin{bmatrix} E_r \\ H_r \end{bmatrix}$$

where for convenience we will choose our plane r to be the junction of the two media and s is a plane within Medium (1), distance l from the junction, $l \leqslant \delta$.

Because of the continuity of the total fields across the junction

$$E_r = E_0 = \frac{J_0}{\sigma_2}$$

$$H_r = \frac{E_0}{\eta_2} = \frac{J_0}{\eta_2 \sigma_2}$$

hence $$J_1 = \left(\frac{\sigma_1}{\sigma_2}\right) J_0 \left[\cosh \gamma_1 l + \frac{\eta_1}{\eta_2} \sinh \gamma_1 l\right]$$

where $$\frac{\eta_1}{\eta_2} = \sqrt{\frac{\mu_1 \sigma_2}{\mu_2 \sigma_1}}$$

Figure 6.5 shows a sketch of the current density in the two media.

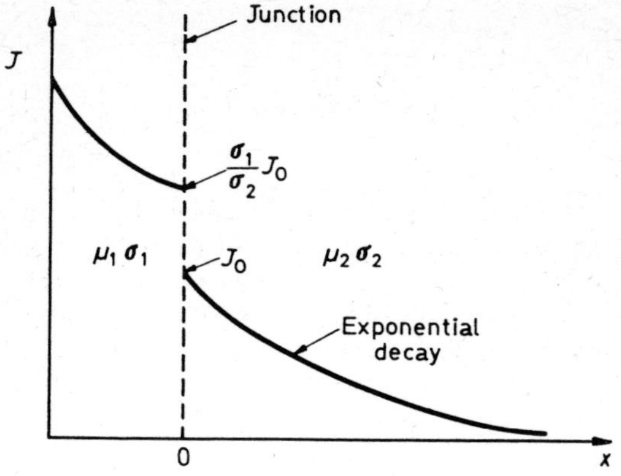

Figure 6.5. The current density in the coating at the junction is $\left(\frac{\sigma_1}{\sigma_2}\right) J_0$ *and so there is a discontinuity across the junction*

CHAPTER 7

7.1. We shall prove the first part of this problem by use of the transverse del-operator. This is the most elegant way of formulating waveguide problems and it is useful for the student to be aware of the technique. It is, of course, not the only method of solving this particular problem.

Maxwell's equations in phasor form are

$$\nabla \times \mathbf{E} = -j\omega\mu \mathbf{H}$$

$$\nabla . \mathbf{E} = 0$$

$$\nabla \times \mathbf{H} = j\omega\epsilon \mathbf{E}$$

$$\nabla . \mathbf{H} = 0$$

Since the conductors are uniform in the z-direction, this is the direction in which propagation will take place. The fields are transverse, that is they lie in the xy-plane and we will therefore write

$$\mathbf{E} = \mathbf{E}_t(x, y) \exp\{-j\beta z\}$$

$$\mathbf{H} = \mathbf{H}_t(x, y) \exp\{-j\beta z\}$$

where the suffix t denotes a transverse vector; a vector with no z-component.

Now let us split the del-operator into two parts, one containing the transverse components of this operator and the other containing the z-component. Thus we write

$$\nabla = \left\{\mathbf{i}\frac{\partial}{\partial x} + \mathbf{j}\frac{\partial}{\partial y}\right\} + \mathbf{k}\frac{\partial}{\partial z} = \nabla_t + \mathbf{k}\frac{\partial}{\partial z}$$

where ∇_t is known as the transverse del-operator.

With these substitutions the curl and divergence of $\mathbf{E}$ become

$$\nabla \times \mathbf{E} = \left\{\nabla_t + \mathbf{k}\frac{\partial}{\partial z}\right\} \times \{\mathbf{E}_t \exp(-j\beta t)\}$$

$$= (\nabla_t \times \mathbf{E}_t) \exp(-j\beta z) - j\beta(\mathbf{k} \times \mathbf{E}_t) \exp(-j\beta z)$$

$$\nabla . \mathbf{E} = (\nabla_t . \mathbf{E}_t) \exp(-j\beta z) - j\beta(\mathbf{k} . \mathbf{E}_t) \exp(-j\beta z) = (\nabla_t . \mathbf{E}_t) \exp(-j\beta z)$$

with a similar pair for $\mathbf{H}$.

Substituting these expressions into Maxwell's equations we find that $\exp(-j\beta z)$ is a common factor and cancels out, a process known as suppressing $\exp(-j\beta z)$. There then results the four equations

$$\nabla_t \times \mathbf{E}_t - j\beta(\mathbf{k} \times \mathbf{E}_t) = -j\omega\mu \mathbf{H}_t$$

$$\nabla_t \cdot \mathbf{E}_t = 0$$

$$\nabla_t \times \mathbf{H}_t - j\beta(\mathbf{k} \times \mathbf{H}_t) = j\omega\epsilon \mathbf{E}_t$$

$$\nabla_t \cdot \mathbf{H}_t = 0$$

Now let us look at the terms on the left of the first and third equations. The student can easily verify that $\nabla_t \times \mathbf{E}_t$ and $\nabla_t \times \mathbf{H}_t$ are both vectors with only z-components. Since the two other terms are vector products they must both be perpendicular to $\mathbf{k}$ and must therefore be transverse vectors. In these two equations, then, we can decompose into z-directed and transverse vectors and equating both sides on this basis results in four equations which, with the two divergence equations are

$$\nabla_t \times \mathbf{E}_t = 0 \qquad \ldots (1)$$

$$\nabla_t \cdot \mathbf{E}_t = 0 \qquad \ldots (2)$$

$$\nabla_t \times \mathbf{H}_t = 0 \qquad \ldots (3)$$

$$\nabla_t \cdot \mathbf{H}_t = 0 \qquad \ldots (4)$$

$$\beta \mathbf{k} \times \mathbf{E}_t = \omega\mu \mathbf{H}_t \qquad \ldots (5)$$

$$\beta \mathbf{k} \times \mathbf{H}_t = -\omega\epsilon \mathbf{E}_t \qquad \ldots (6)$$

Thus $\mathbf{E}_t$ can be calculated from the two equations

$$\nabla_t \times \mathbf{E}_t = 0$$

$$\nabla_t \cdot \mathbf{E}_t = 0$$

which are the two-dimensional electrostatic equations. Similarly $\mathbf{H}_t$ can be calculated from the two-dimensional magnetostatic equations

$$\nabla_t \times \mathbf{H}_t = 0$$

$$\nabla_t \cdot \mathbf{H}_t = 0$$

This proves the first part of the question.

Now, from Equations (5) and (6), which only exist if the fields are time-varying and which link together the electric and magnetic fields, we have, premultiplying the first by k x.

$$\beta \mathbf{k} \times (\mathbf{k} \times \mathbf{E}_t) = \omega\mu \mathbf{k} \times \mathbf{H}_t = -\frac{\omega^2 \mu\epsilon}{\beta}$$

and since

$$\mathbf{k} \times (\mathbf{k} \times \mathbf{E}_t) = -\mathbf{E}_t$$

$$\therefore \qquad \beta^2 = \omega^2 \mu\epsilon$$

Letting

$$\beta = +\omega\sqrt{\mu\epsilon}$$

we obtain the two equations (which are in fact the same equation)

$$\mathbf{H}_t = \sqrt{\frac{\epsilon}{\mu}}\,(\mathbf{k} \times \mathbf{E}_t)$$

$$\mathbf{E}_t = -\sqrt{\frac{\mu}{\epsilon}}\,(\mathbf{k} \times \mathbf{H}_t)$$

as the equations which determine $\mathbf{H}_t$ from $\mathbf{E}_t$ or vice versa.

The paradox arises because we have been considering waves to flow in one direction only, in the above this direction was the direction of positive z. If we take $\mathbf{E}_t$ as a measure of the voltage on the line and $\mathbf{H}_t$ as a measure of the current then there would be a definite ratio between the two.

Now a wave in the direction of negative z is obtained simply by letting β take the opposite sign. This not only changes the sign in the exponential $\exp(-j\beta z)$ but if we look back to the set of six equations we see that it will change the sign of $\mathbf{H}_t$ relative to $\mathbf{E}_t$. This means that the linking equation for $\mathbf{H}_t$ becomes

$$\mathbf{H}_t = -\sqrt{\frac{\epsilon}{\mu}}\,(\mathbf{k} \times \mathbf{E}_t)$$

Now let us add a forward wave denoted by $\mathbf{E}_{t_1}$ to a backward wave denoted by $\mathbf{E}_{t_2}$. The total field $\mathbf{E}_t$ will be given by

$$\mathbf{E}_t = [\mathbf{E}_{t_1} \exp(-j\beta z) + \mathbf{E}_{t_2} \exp(+j\beta z)]$$

and the total $\mathbf{H}_t$ (using the linking equations which are appropriate to the propagation direction)

$$\mathbf{H}_t = \sqrt{\frac{\epsilon}{\mu}}\ \mathbf{k} \times [\mathbf{E}_{t_1} \exp(-\mathrm{j}\beta z) - \mathbf{E}_{t_2} \exp(+\mathrm{j}\beta z)]$$

Since $\mathbf{E}_{t_1}$ and $\mathbf{E}_{t_2}$ can have an arbitrary relation the negative sign in the bracket for $\mathbf{H}_t$ means that the ratio of $\mathbf{E}_t$ to $\mathbf{H}_t$ is now no longer fixed and will in fact take on a value determined by the load. The paradox is therefore resolved.

7.2. Again a problem where the transverse del-operator can be used to advantage.

Maxwell's equations in phasor form are

$$\nabla \times \mathbf{E} = -\mathrm{j}\omega\mu\mathbf{H}$$

$$\nabla \times \mathbf{H} = \mathrm{j}\omega\epsilon\mathbf{E}$$

Now let

$$\mathbf{E} = (\mathbf{E}_t + \mathbf{k}e_z) \exp(-\mathrm{j}\beta z)$$

$$\mathbf{H} = \mathbf{H}_t \exp(-\mathrm{j}\beta z)$$

and

$$\nabla = \nabla_t + \mathbf{k}\frac{\partial}{\partial z}$$

where we have dropped the exponential time dependence which is actually used in the problem. (This can be replaced at any time.)

If these equations are substituted into Maxwell's equations and $\exp(-\mathrm{j}\beta z)$ suppressed as in the previous problem we obtain

$$\nabla_t \times \mathbf{E}_t - \mathrm{j}\beta\mathbf{k} \times \mathbf{E}_t + \nabla_t \times (\mathbf{k}e_z) - \mathrm{j}\beta\mathbf{k} \times \mathbf{k}e_z = -\mathrm{j}\omega\mu\mathbf{H}_t$$

$$\nabla_t \times \mathbf{H}_t - \mathrm{j}\beta\mathbf{k} \times \mathbf{H}_t = \mathrm{j}\omega\epsilon\mathbf{E}_t + \mathrm{j}\omega\epsilon\mathbf{k}e_z$$

These equations are split up into z-directed and transverse components as before to obtain (note $\mathbf{k} \times \mathbf{k} = 0$) the four equations

$$\nabla_t \times \mathbf{E}_t = 0$$

$$\nabla_t \times \mathbf{H}_t = \mathrm{j}\omega\epsilon\mathbf{k}e_z$$

$$\mathrm{j}\beta\mathbf{k} \times \mathbf{E}_t - \mathrm{j}\omega\mu\mathbf{H}_t = \nabla_t \times (\mathbf{k}e_z)$$

$$-\mathrm{j}\beta\mathbf{k} \times \mathbf{H}_t = \mathrm{j}\omega\epsilon\mathbf{E}_t$$

From the fourth

$$\mathbf{E}_t = -\frac{\beta}{\omega\epsilon}\mathbf{k} \times \mathbf{H}_t$$

whence

$$\mathbf{k} \times \mathbf{E}_t = \frac{\beta}{\omega\epsilon}\mathbf{H}_t$$

and substituting into the third we obtain an expression for $\mathbf{H}_t$ in terms of e_z.

$$\mathbf{H}_t = \frac{\mathrm{j}\omega\epsilon}{(\omega^2\mu\epsilon - \beta^2)}\nabla_t \times (\mathbf{k}e_z)$$

giving $\mathbf{E}_t$ in terms of e_z as

$$\mathbf{E}_t = -\frac{\mathrm{j}\beta}{(\omega^2\mu\epsilon - \beta^2)}\mathbf{k} \times [\nabla_t \times (\mathbf{k}e_z)]$$

The whole field can therefore be derived from e_z (or E_z) as the problem requires us to show.

For the field given in the problem the e_z we are using is just

$$e_z = E_0 \sin\frac{\pi x}{x_0}\sin\frac{\pi y}{y_0}$$

Thus

$$\nabla_t \times (\mathbf{k}e_z) = \mathbf{i}\frac{\pi E_0}{y_0}\sin\frac{\pi x}{x_0}\cos\frac{\pi y}{y_0} - \mathbf{j}\frac{\pi E_0}{x_0}\cos\frac{\pi x}{x_0}\sin\frac{\pi y}{y_0}$$

and

$$\mathbf{k} \times [\nabla_t \times (\mathbf{k}e_t)] = \mathbf{i}\frac{\pi E_0}{x_0}\cos\frac{\pi x}{x_0}\sin\frac{\pi y}{y_0} + \mathbf{j}\frac{\pi E_0}{y_0}\sin\frac{\pi x}{x_0}\cos\frac{\pi y}{y_0}$$

If $\mathbf{E}_t$ and $\mathbf{H}_t$ are now calculated and the exponential terms added as the problem indicates we obtain the expressions for all the other field components

$$E_x = -\frac{\mathrm{j}\beta}{(\omega^2\mu\epsilon - \beta^2)}\frac{\pi E_0}{x_0}\cos\frac{\pi x}{x_0}\sin\frac{\pi y}{y_0}\exp \mathrm{j}\,(\omega t - \beta z)$$

$$E_y = -\frac{\mathrm{j}\beta}{(\omega^2\mu\epsilon - \beta^2)}\frac{\pi E_0}{y_0}\sin\frac{\pi x}{x_0}\cos\frac{\pi y}{y_0}\exp \mathrm{j}\,(\omega t - \beta z)$$

$$H_x = \frac{\mathrm{j}\omega\epsilon}{(\omega^2\mu\epsilon - \beta^2)}\frac{\pi E_0}{y_0}\sin\frac{\pi x}{x_0}\cos\frac{\pi y}{y_0}\exp \mathrm{j}\,(\omega t - \beta z)$$

$$H_y = -\frac{\mathrm{j}\omega\epsilon}{(\omega^2\mu\epsilon - \beta^2)}\frac{\pi E_0}{x_0}\cos\frac{\pi x}{x_0}\sin\frac{\pi y}{y_0}\exp \mathrm{j}\,(\omega t - \beta z)$$

On any conducting walls E_z (or e_z) must be zero. This means that $(\pi x/x_0)$ and $(\pi y/y_0)$ must be an integral number of π radians. Thus the conducting walls must be placed on the planes

$$x = mx_0 \quad \text{where } m \text{ is any integer}$$

and

$$y = ny_0 \quad \text{where } n \text{ is any integer.}$$

The student should check that the boundary conditions for the other components are not violated at these planes.

The smallest waveguide is clearly obtained when two planes for which m changes by unity are combined with two planes for which n changes by unity. This means that the smallest waveguide has a rectangular cross-section, one dimension being x_0 and the other being y_0.

If the expression for $\mathbf{H}_t$ in terms of e_z is substituted into the equation

$$\boldsymbol{\nabla}_t \times \mathbf{H}_t = \mathrm{j}\omega\epsilon\mathbf{k}e_z$$

we obtain

$$\boldsymbol{\nabla}_t \times (\boldsymbol{\nabla}_t \times \mathbf{k}e_z) = (\omega^2\mu\epsilon - \beta^2)\mathbf{k}e_z$$

or

$$\nabla_t{}^2 e_z + (\omega^2\mu\epsilon - \beta^2)e_z = 0$$

since

$$\boldsymbol{\nabla}_t \cdot (\mathbf{k}e_z) \equiv 0$$

Now, using the known expression for e_z it is easily shown that

$$\nabla_t^2 e_z = -\left[\left(\frac{\pi}{x_0}\right)^2 + \left(\frac{\pi}{y_0}\right)^2\right] e_z$$

and therefore

$$\beta^2 = \omega^2\mu\epsilon - \left(\frac{\pi}{x_0}\right)^2 - \left(\frac{\pi}{y_0}\right)^2$$

For propagation the phase constant β must be real and therefore $\beta^2 \geqslant 0$. Thus the frequency ω must satisfy

$$\omega^2\mu\epsilon \geqslant \left(\frac{\pi}{x_0}\right)^2 + \left(\frac{\pi}{y_0}\right)^2$$

or, rewriting in terms of f_c where f_c is the first frequency at which the equation is satisfied then

$$f_c = \frac{c}{2}\left\{\frac{1}{x_0^2} + \frac{1}{y_0^2}\right\}^{1/2}$$

where c is the velocity of light.

7.3. Consider the boundary condition on E_x. Since this component must be zero on any tangential conducting wall it must vanish when $y = \pm b$.

Thus $$\cos(kb \sin\theta) = 0$$

This means that θ is related to the free space wavelength (since $k = 2\pi/\lambda$) by the equation

$$\lambda = 4b \sin\theta$$

Now the phase constant for propagation $\beta = k \cos\theta$ and the cut-off condition is obtained when this phase constant is zero, that is when $\theta = \pi/2$. The free space wavelength at this condition, which is the cut-off wavelength, is therefore given by

$$\lambda_c = 4b$$

Above cut-off the guide wavelength is related to β by $\beta = 2\pi/\lambda_g$ and we therefore obtain the two equations

$$k \cos\theta = 2\pi/\lambda_g$$

and
$$k \sin\theta = \pi/2b = 2\pi/\lambda_c$$

If θ is now eliminated between these two equations we obtain

$$k^2 = \left(\frac{2\pi}{\lambda_g}\right)^2 + \left(\frac{2\pi}{\lambda_c}\right)^2$$

whence

$$\frac{1}{\lambda_g{}^2} = \frac{1}{\lambda^2} - \frac{1}{\lambda_c^2}$$

which is the relationship required.

The easiest way to assess whether or not slits in waveguide walls will disturb the waves is to consider the direction of surface current flow on the inside of the walls. If the current flows along the length of a narrow slit the waves will not be disturbed whereas if the current flow has a component *across* the slit there will be a considerable effect. If this argument is applied to the problem in hand, on the planes $y = \pm b$ the current density vector $\mathbf{J} = \mathbf{n} \times \mathbf{H} = \pm \mathbf{i}H_z$. The surface current, therefore, flows in the x-direction (positive or negative) only. Thus slits parallel to the x-axis will not disturb the waves whereas slits parallel to the z-axis can be expected to have a considerable effect.

The total time-average power flow along the waveguide is obtained by calculating the real part of the z-component of the complex Poynting vector and integrating this over the cross-section of the waveguide.

In our case the z-component of the complex Poynting vector is

$$\mathbf{k} \,.\, \mathbf{S} = \tfrac{1}{2}\mathbf{k} \,.\, \mathbf{E} \times \mathbf{H}^* = \tfrac{1}{2}\mathbf{k} \,.\, (-\mathbf{j}E_x H_z^* + \mathbf{k}E_x H_y^*) = \tfrac{1}{2} E_x H_y^*$$

Inserting the fields given we obtain

$$\mathbf{k} \,.\, \mathbf{S} = \frac{1}{2}\sqrt{\frac{\mu}{\epsilon}}\,|H_0|^2 \cos\theta \cos^2(ky \sin\theta)$$

where we have used the relation between E_0 and H_0.

This quantity is itself real and can therefore be integrated over the cross-section of the waveguide directly. The integrations are trivial and yield

$$\text{Time-average power flow} = \frac{1}{2}\sqrt{\frac{\mu}{\epsilon}}\,|H_0|^2 \cos\theta \int_{x=-a}^{+a}\int_{y=-b}^{+b} \cos^2(ky\sin\theta)\,\mathrm{d}y\,\mathrm{d}x$$

$$= ab\sqrt{\frac{\mu}{\epsilon}}\,|H_0|^2\cos\theta$$

7.4. This problem really asks for the first fundamental calculations which must be carried out for any new waveguide.

The clearest analysis probably arises with the use of the transverse del-operator introduced in Problem 7.1 and used for a waveguide problem in 7.2. The details of the analysis follow closely those of Problem 7.2 except that in this case we are dealing with a TE mode instead of a TM mode. We therefore define

$$\mathbf{E} = \mathbf{E}_t \exp(-\mathrm{j}\beta z)$$

$$\mathbf{H} = (\mathbf{H}_t + \mathbf{k}h_z)\exp(-\mathrm{j}\beta z)$$

and

$$\boldsymbol{\nabla} = \boldsymbol{\nabla}_t + \mathbf{k}\frac{\partial}{\partial z}$$

These are substituted into Maxwell's equations and the result split into transverse and z-directed components. We will not repeat the detailed algebra here since it is so similar to that of Problem 7.2. From the decomposition we obtain four equations as before and use two of them to express the transverse fields in terms of h_z. The results of this are

$$E_t = -\frac{\mathrm{j}\omega\mu}{(\omega^2\mu\epsilon - \beta^2)}\boldsymbol{\nabla}_t \times (\mathbf{k}h_z)$$

$$H_t = -\frac{\mathrm{j}\beta}{(\omega^2\mu\epsilon - \beta^2)}\mathbf{k}\times[\boldsymbol{\nabla}_t\times(\mathbf{k}h_z)]$$

One of the two remaining equations is

$$\boldsymbol{\nabla}_t \times \mathbf{E}_t = -\mathrm{j}\omega\mu\mathbf{k}h_z$$

and if the above expression for $\mathbf{E}_t$ is substituted we obtain the differential equation for h_z as

$$\nabla_t^2 h_z + (\omega^2 \mu \epsilon - \beta^2) h_z = 0$$

This equation now has to be solved for the particular conditions given, that is, a circular waveguide and a field which is independent of the angular coordinate. The former fact means that we should work in cylindrical co-ordinates and the latter means that h_z is a function of r alone. Thus

$$\nabla_t^2 h_z = \frac{1}{r}\frac{\mathrm{d}}{\mathrm{d}r}\left(r\frac{\mathrm{d}h_z}{\mathrm{d}r}\right)$$

and if we denote

$$k_c^2 = \omega^2 \mu \epsilon - \beta^2$$

the differential equation satisfied by h_z is

$$\frac{1}{r}\frac{\mathrm{d}}{\mathrm{d}r}\left(r\frac{\mathrm{d}h_z}{\mathrm{d}r}\right) + k_c^2 h_z = 0$$

or

$$\frac{\mathrm{d}^2 h_z}{\mathrm{d}r^2} + \frac{1}{r}\frac{\mathrm{d}h_z}{\mathrm{d}r} + k_c^2 h_z = 0$$

This equation is a standard form with the known solution

$$h_z = AJ_0(k_c r)$$

where A is an arbitrary constant and $J_0(x)$ is the zero order Bessel function.

(It should be noted by the student that the other solution to the second order Bessel equation $N_0(k_c r)$ is infinite at $r = 0$ and since a circular waveguide includes the region $r = 0$ the arbitrary constant multiplying this solution must be put equal to zero. We have therefore ignored it.) If this solution for h_z is inserted into the transverse field equations we immediately find that $\mathbf{E}_t$ has an angular component E_ϕ only and $\mathbf{H}_t$ has a radial component H_r only. Carrying out the operations the field components are then

$$H_z = AJ_0(k_c r) \exp(-j\beta z)$$

$$E_\phi = jA\frac{\omega\mu}{k_c} J_0'(k_c r) \exp(-j\beta z)$$

$$H_r = -jA\frac{\beta}{k_c} J_0'(k_c r) \exp(-j\beta z)$$

In these equations the notation $J_0'(k_c r)$ means $[\mathrm{d}\,J(x)/\mathrm{d}x]$ with x put equal to $k_c r$. It does not mean $[\mathrm{d}J(k_c r)/\mathrm{d}r]$. There now remains the question of the boundary conditions. Clearly E_ϕ (or H_r) must be zero on the conducting wall, that is when $r = a$. Thus the at present unknown parameter k_c (which is only defined in terms of β at present) must satisfy the equation

$$J_0'(k_c a) = 0$$

Now this equation has many roots. At cut-off, however, $\beta = 0$ and therefore k_c is directly proportional to the cut-off frequency. Since the problem asks for the lowest mode we seek the lowest value of cut-off frequency and hence the lowest value of k_c. This means that we must take the lowest root of the above equation which is

$$k_c a = 3{\cdot}832$$

an expression which determines k_c and also β.

In order to calculate the attenuation of the waveguide for this mode we must compare the power lost in the walls with the power transmitted down the guide. Turning our attention to the power transmitted down the guide this can be computed from the complex Poynting vector by taking the real part of the z-component. Using the expressions for the fields we find

$$\begin{aligned}\mathbf{k}\,.\,\mathbf{S} &= -\tfrac{1}{2} E_\phi H_r^* \\ &= \frac{\omega\mu\beta}{2k_c^2} |A|^2 \{J_0'(k_c r)\}^2\end{aligned}$$

and thus the power transmitted, P_t, is given by

$$P_t = \frac{\omega\mu\beta}{2k_c^2}|A|^2 \int_0^a 2\pi r\{J_0'(k_c r)\}^2 \mathrm{d}r = \frac{\pi\omega\mu\beta}{k_c^2}|A|^2 \int_0^a r\{J_0'(k_c r)\}^2 \mathrm{d}r$$

Now we come to the calculation of the power absorbed in the walls; in fact we need the power lost per unit length. The evaluation of this quantity assumes that the attenuation is small, or to put it another way, that the effect of the imperfectly conducting walls is negligible on the waveguide fields, in particular on the magnetic field. We assume that the only effect is the existence of a small component of electric field tangential to the wall. In Problems **6.6** and **6.7** we have already met the concept of wave impedance of an imperfect conductor which we shall denote here by Z_m and which is calculated from

$$Z_m = \sqrt{\frac{\mathrm{j}\omega\mu}{\sigma}}$$

Now at the wall (in this case) only the z-component of magnetic field H_z is non-zero. If we assume that the small tangential electric field and the tangential magnetic field constitute a plane wave (strictly in this case a cylindrical wave) travelling into the wall then the electric field $E_{\tan}$ is related to the magnetic field by

$$E_{\tan} = Z_m H_z$$

and is, of course in a direction perpendicular to H_z. We can now calculate the loss in the wall by calculating the component of the complex Poynting vector *into* the wall. This can then be integrated over a unit length to find the power lost per unit length.

Since, at the wall,

$$H_z = AJ_0(k_c a) \exp(-\mathrm{j}\beta z)$$

then

$$E_{\tan} = Z_m AJ_0(k_c a) \exp(-\mathrm{j}\beta z)$$

The component of the complex Poynting vector *into* the wall is then

$$\tfrac{1}{2} E_{\tan} H_z^* = \tfrac{1}{2} |A|^2 Z_m J_0^2(k_c a)$$

Integrating the real part of this over the surface of the wall of unit length gives the power lost per unit length P_l as

$$P_l = 2\pi a \times \tfrac{1}{2} |A|^2 J_0^2(k_c a)(\text{Re } Z_m) = \sqrt{\frac{\omega\mu}{2\sigma}}\ \pi a |A|^2 J_0^2(k_c a)$$

Now, if the attenuation coefficient α is in the units of nepers/unit length then the power transmitted down the guide, including attenuation is proportional to exp $(-2\alpha z)$. The power lost in a length dz is therefore just (-2α) times the power flowing down the guide as this point. Thus

$$P_l \, \mathrm{d}z = 2\alpha P_t \, \mathrm{d}z$$

or

$$\alpha = \frac{P_l}{2P_t}$$

Before putting in the values we should evaluate the integral which was left in the calculation of P_t. (This was left because it is a purely mathematical point and is not part of the discussion.) Since

$$J_0'(x) = -J_1(x)$$

the integral is

$$\int_0^a r J_1^2(k_c r) \, \mathrm{d}r$$

As far as Bessel functions are concerned this is a standard integral with the result, in general,

$$\int_0^a r J_1^2(k_c r) \, \mathrm{d}r = \frac{a^2}{2}\left\{ [J_1'(k_c a)]^2 + \left(1 - \frac{1}{k_c^2 a^2}\right) J_1^2(k_c a)\right\}$$

In our case, however, $J_1(k_c a)$, or $-J_0'(k_c a)$, is zero, a property which arose out of the boundary conditions. Further, at this value the quantity $[J_1'(k_c a)]^2$ is equal to $[J_0(k_c a)]^2$. Thus our particular integral is equal to

$$\frac{a^2}{2} J_0^2(k_c a)$$

Inserting all the relevant data into the expression for the attenuation coefficient we finally obtain

$$\alpha = \frac{1}{\sqrt{2\omega\mu\sigma}} \cdot \frac{k_c^2}{\beta a}$$

7.5. In the first place, this mode, at the cylindrical walls of a cavity resonator, has a *z*-component of magnetic field only, as was shown in the previous problem. The current flow in the walls is therefore circumferentially around the walls and there is no requirement for the current to flow from the cylindrical walls to the flat end walls of the resonator. This means that a small gap between the end and side walls as shown in *Figure 7.4* does

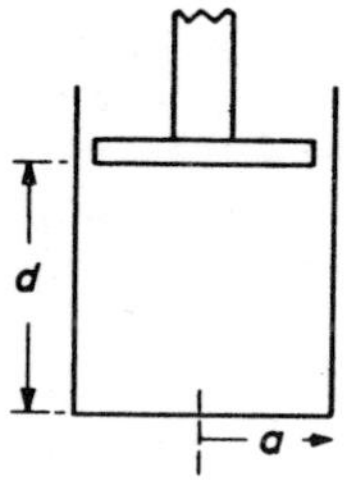

Figure 7.4.

not interrupt the lines of current flow and therefore will not perturb the fields of the resonant mode under discussion. Should, in fact, any of the more dominant modes of a cylindrical waveguide be excited by the coupling element they will be heavily affected and their fields will penetrate behind the end wall. If an absorbent medium is placed behind this end wall, therefore, it will tend to absorb much of the power in these modes. This process does represent an unwanted loss of power and the coupling elements should be designed as far as possible to excite only the desired mode.

In considering the coupling to and from this mode of resonance we must consider the fields of all the modes of a cylindrical waveguide which have a lower or equal cut-off frequency. These are, in order of the cut-off frequency,

$$TE_{11}, TM_{01}, TE_{21}, TM_{11} = TE_{01}$$

the last being the wanted mode.

An exceedingly dangerous mode for resonant cavity use is the TM_{11} which, having the same cut-off frequency, will have the same guide wavelength and the cavity will therefore have the same resonant frequency. Fortunately neither this mode nor the TM_{01} mode has a longitudinal component of the magnetic field and if the coupling elements excite *only* such a component then neither of the transverse magnetic modes can be excited at the outset. This points to a coupling element which is in the form of a small loop, the plane of which is parallel to the end plates of the cavity, or a small hole coupled to a feeding waveguide in which the waveguide transmission mode has a component of magnetic field (at its wall) only in the direction of the axis of the cylinder.

Either of these methods could, in principle, excite the TE_{11} and TE_{21} modes in addition to the desired TE_{01} mode. Since the resonant frequencies of these modes will be different if the space behind the movable end plate does not allow a resonance of the unwanted modes at the same frequency as that of the wanted mode, then the amplitude of excitation will be small and little loss will result from their presence. They will, of course, resonate at other frequencies and will limit the frequency excursion of the tunable system. Notice that if the resonator has input and output loops spaced 90° apart around the circumference then the output loop will not couple to any TE_{11} mode set up by the input loop since there is no magnetic field at this position.

The fields of the resonant modes of cavities such as this can easily be calculated from the corresponding fields of the waveguide modes. The resonant field is just a standing wave field in the waveguide and can therefore be decomposed into two travelling waves in opposite directions.

First of all let us repeat the fields of the TE_{01} mode as found in Problem 7.4.

$$H_z = AJ_0(k_c r) \exp(-j\beta z)$$

$$E_\phi = jA \frac{\omega\mu}{k_c} J_0'(k_c r) \exp(-j\beta z)$$

$$H_r = -jA \frac{\beta}{k_c} J_0'(k_c r) \exp(-j\beta z)$$

These fields are written for a wave travelling in the direction of positive z. In order to obtain a wave travelling in the other direction all we have to do is to change the sign of β wherever it occurs. Obviously the exponents will change sign, but so will the expression for H_r. Thus the fields for the wave travelling in the direction of negative z are

$$H_z = AJ_0(k_c r) \exp(+j\beta z)$$

$$E_\phi = jA \frac{\omega\mu}{k_c} J_0'(k_c r) \exp(+j\beta z)$$

$$H_r = jA \frac{\beta}{k_c} J_0'(k_c r) \exp(+j\beta z)$$

The amplitudes, A, are arbitrary in these expressions.

Suppose that we start to construct the cavity by first of all placing a conducting plate at $z = 0$. This means that at $z = 0$ the total tangential electric field must be zero. We therefore have to choose the relative amplitudes to satisfy this requirement. A glance at the field equations will suffice to show that if the amplitude of the positive going wave is A then the amplitude of the negative going wave must be $-A$. If $-A$ is substituted for A in the second set and the two sets combined we will obtain a field which satisfies the boundary condition imposed by a conducting plate at $z = 0$. The resultant field is

$$H_z = -2jAJ_0(k_c r) \sin \beta z$$

$$E_\phi = 2A \frac{\omega\mu}{k_c} J_0'(k_c r) \sin \beta z$$

$$H_r = -2jA \frac{\beta}{k_c} J_0'(k_c r) \cos \beta z$$

Now let us close the cavity by placing a conducting plate at $z = d$. Again the electric field must vanish which can only happen if

$$\sin \beta d = 0$$

or

$$\beta d = n\pi$$

Since β is, apart from the radius, a function of frequency alone this requirement determines the frequency at which such a field can exist and this is the resonant frequency of the cavity. For the first resonant mode we must choose the value of n which gives the lowest frequency, i.e. $n = 1$. The final field expressions for this mode, the TE_{011}, are therefore

$$H_z = -2jAJ_0(k_c r)\sin\left(\frac{\pi z}{d}\right)$$

$$E_\phi = 2A\frac{\omega\mu}{k_c} J_0'(k_c r)\sin\left(\frac{\pi z}{d}\right)$$

$$H_r = -2jA\frac{\pi}{k_c d} J_0'(k_c r)\cos\left(\frac{\pi z}{d}\right)$$

where $$\omega^2\mu\epsilon = k_c^2 + \left(\frac{\pi}{d}\right)^2$$

The Q of resonant cavities is calculated from the definition

$$Q = \frac{\omega\text{ (time-average stored energy)}}{\text{(energy loss per second)}} = \frac{\omega W}{P}$$

where ω is the resonant frequency.

Now it is readily shown (and the student should satisfy himself that it is so) that at resonance the time-average stored energy in the electric field is the same as that in the magnetic field and we need to do only one calculation. Thus

$$W = 2\int_v \frac{1}{4}\epsilon|E|^2\,dV = \frac{\epsilon}{2}\int_v |E|^2\,dV$$

$$= 2\epsilon|A|^2\left(\frac{\omega\mu}{k_c}\right)^2\int_{z=0}^{d}\sin^2\left(\frac{\pi z}{d}\right)dz\int_{r=0}^{a} 2\pi r\{J_0'(k_c r)\}^2 dr$$

$$= 2\pi\epsilon d|A|^2\left(\frac{\omega\mu}{k_c}\right)^2\int_0^a r\{J_0'(k_c r)\}^2\,dr$$

The integral is the same as that met in the previous problem and hence we finally obtain

$$W = \epsilon\pi a^2 d|A|^2\left(\frac{\omega\mu}{k_c}\right)^2 J_0^2(k_c a)$$

The calculation of the energy loss is essentially the same as that of the attenuation in a waveguide except that loss in the end walls must also be considered.

On the cylindrical walls the magnetic field is

$$H_z = -2jA\,J_0(k_c a)\sin\left(\frac{\pi z}{d}\right)$$

giving rise to a tangential electric field

$$\mathbf{E}_{\tan} = -2jA\sqrt{\frac{j\omega\mu}{\sigma}}\,J_0(k_c a)\sin\left(\frac{\pi z}{d}\right)\mathbf{a}_\phi$$

Integration of the real part of the complex Poynting vector (the component into the walls, of course) gives us the power lost in the cylindrical walls as

$$P_1 = \int \mathrm{Re}\{\tfrac{1}{2}\mathbf{a}_r \,.\, \mathbf{E}_{\tan} \times \mathbf{H}^*\}\,\mathrm{d}s$$

$$= \sqrt{\frac{\omega\mu}{2\sigma}}\,2|A|^2 J_0^2(k_c a)\int_{z=0}^{d} 2\pi a\sin^2\left(\frac{\pi z}{d}\right)\mathrm{d}z$$

$$= \sqrt{\frac{\omega\mu}{2\sigma}}\,2\pi a d|A|^2 J_0{}^2(k_c a)$$

On the end walls the magnetic field is (for the wall $z = 0$)

$$H_r = -2jA\frac{\pi}{k_c d}J_0'(k_c r)$$

giving rise to a tangential electric field

$$\mathbf{E}_{\tan} = -2jA\sqrt{\frac{j\omega\mu}{\sigma}}\left(\frac{\pi}{k_c d}\right)J_0'(k_c r)\,\mathbf{a}_\phi$$

The power lost in this wall is therefore

$$\sqrt{\frac{\omega\mu}{2\sigma}}\,4\pi|A|^2\left(\frac{\pi}{k_c d}\right)^2\int_0^a r\{J_0'(k_c r)\}^2\,\mathrm{d}r$$

and in both end walls

$$P_2 = \sqrt{\frac{\omega\mu}{2\sigma}}\, 4\pi|A|^2 \left(\frac{\pi a}{k_c d}\right)^2 J_0^2(k_c a)$$

We finally obtain the value of the Q as

$$Q = \frac{\omega W}{P_1 + P_2}$$

$$= \sqrt{\frac{\omega\mu\sigma}{2}} \cdot \left\{\frac{k^2 a d^3}{k_c^2 d^3 + 2\pi^2 a}\right\}$$

where $$k^2 = \omega^2\mu\epsilon$$

7.6. From Problem 7.1 we know that the fields can be calculated from the electrostatic equations. Hence a simple calculation will show that the electric field is radial and can be written as

$$E_r = \frac{A}{r} \exp(-j\beta z)$$

where $$\beta^2 = \omega^2\mu\epsilon$$

Since (also from Problem 7.1)

$$\mathbf{H}_t = \sqrt{\frac{\epsilon}{\mu}}\, \mathbf{k} \times \mathbf{E}_t$$

we obtain

$$H_\phi = \frac{A}{r}\sqrt{\frac{\epsilon}{\mu}} \exp(-j\beta z)$$

and these equations give the fields in the coaxial line.

The calculation of attenuation coefficient must be carried out in two parts, one for each conductor. On the inner conductor surface

$$H_\phi = \frac{A}{a}\sqrt{\frac{\epsilon}{\mu}}\exp(-\mathrm{j}\beta z)$$

resulting in a tangential component of electric field

$$\mathbf{E} = \sqrt{\frac{\mathrm{j}\omega\mu}{\sigma}}\,\mathbf{a}_r \times \mathbf{H} = \sqrt{\frac{\mathrm{j}\omega\mu}{\sigma}}\,H_\phi\,\mathbf{a}_z$$

Carrying out the usual calculations on the complex Poynting vector the loss per unit length in this wall is just

$$\int \mathrm{Re}[-\tfrac{1}{2}\,\mathbf{a}_r \cdot \mathbf{E} \times \mathbf{H}^*]\,\mathrm{d}S = \sqrt{\frac{\omega\mu}{2\sigma}}\left(\frac{\epsilon}{\mu}\right)|A|^2 \cdot \frac{\pi}{a}$$

In a similar manner the loss in the outer wall is

$$\sqrt{\frac{\omega\mu}{2\sigma}}\left(\frac{\epsilon}{\mu}\right)|A|^2\frac{\pi}{b}$$

giving the total power loss per unit length as

$$P_l = \sqrt{\frac{\omega\mu}{2\sigma}}\left(\frac{\epsilon}{\mu}\right)|A|^2\pi\left(\frac{1}{a}+\frac{1}{b}\right)$$

The power flow down the line for this field is

$$P_t = \int_{r=a}^{b}\{\mathrm{Re}\tfrac{1}{2}\,\mathbf{a}_z \cdot \mathbf{E} \times \mathbf{H}^*\}\,2\pi r\,\mathrm{d}r = \pi|A|^2\sqrt{\frac{\epsilon}{\mu}}\ln\left(\frac{b}{a}\right)$$

and the attenuation coefficient (following the method in Problem 7.4) is

$$\alpha = \frac{P_l}{2P_t} = \sqrt{\frac{\epsilon}{\mu}}\sqrt{\frac{\omega\mu}{2\sigma}}\left\{\frac{\dfrac{1}{a}+\dfrac{1}{b}}{\ln(b/a)}\right\}\ \text{nepers/m}$$

Since we have already calculated the power flow P_t, the characteristic impedance is most readily found from this and the total longitudinal current, I, using

$$Z_0 = 2P_t/|I|^2$$

On the inner conductor the surface current density $\mathbf{J}_s$ is

$$\mathbf{J}_s = \mathbf{a}_r \times \mathbf{H} = \mathbf{a}_z (H_\phi)_{r=a}$$

whence

$$J_z = \frac{A}{a}\sqrt{\frac{\epsilon}{\mu}} \exp(-\mathrm{j}\beta z)$$

and the total current, I, is

$$I = 2\pi a J_z = 2\pi A \sqrt{\frac{\epsilon}{\mu}} \exp(-\mathrm{j}\beta z)$$

giving the characteristic impedance

$$Z_0 = \frac{1}{2\pi}\sqrt{\frac{\mu}{\epsilon}} \ln(b/a)$$

The remaining parts of the question require a minimization procedure. (*a*) If b is fixed at b_0 say, the attenuation coefficient is proportional to

$$\frac{1 + \left(\dfrac{b_0}{a}\right)}{\ln\left(\dfrac{b_0}{a}\right)} = \frac{1 + x}{\ln x}$$

where we have put $x = (b_0/a)$

By differentiating we can show that this is a minimum with x if

$$x \ln x = 1 + x$$

which has the solution

$$x = 3{\cdot}6$$

whence the characteristic impedance for air spaced line with the lowest attenuation coefficient is

$$Z_0 = \frac{1}{2\pi} \times 120\pi \times \ln\ 3{\cdot}6 = 77\ \Omega$$

(*b*) The maximum power which can be passed down the line will be that power for which the maximum value of the electric field (in this case at the surface of the inner conductor) is equal to the breakdown field strength of the dielectric, air. Denote this breakdown field strength by E_b. Then since the maximum value of the electric field is $|A|/a$, we find

$$|A| = aE_b$$

and the maximum power which can flow down a given line is

$$P = \pi E_b^2 \sqrt{\frac{\epsilon}{\mu}}\, a^2 \ln\left(\frac{b}{a}\right)$$

Now, since $b = b_0$ is fixed, we can write

$$P \propto \frac{\ln x}{x^2}$$

and, differentiating, find that this is a maximum if

$$2 \ln x = 1 \text{ or } x = \sqrt{e}$$

giving a value of characteristic impedance

$$Z_0 = \frac{1}{2\pi} \times 120\pi \times \frac{1}{2} = 30\ \Omega$$

7.7. In order to solve this problem we require the fields of the TE_{10} mode in the propagating region and in the evanescent region. First of all we use our knowledge of the incident wave to say that:

(*i*) the only field components existing will be E_y, H_x and H_z (referring the waveguide cross-section to the axes shown in *Figure 7.5*),

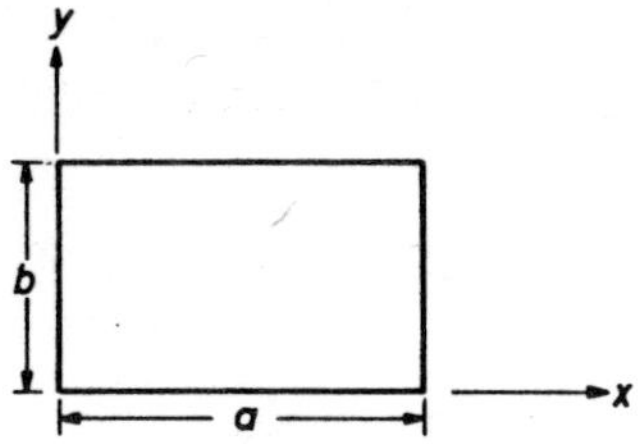

Figure 7.5.

(*ii*) in both regions the dependence on x will be of the form sin $(\pi x/a)$. Thus we can define E_y of the form

$$E_y = A \sin\left(\frac{\pi x}{a}\right) \exp(\gamma z)$$

and since $\nabla \times \mathbf{E} = -j\omega\mu\mathbf{H}$, readily derive

$$H_x = \frac{\gamma}{j\omega\mu} A \sin\left(\frac{\pi x}{a}\right) \exp(\gamma z)$$

$$H_z = -\frac{(\pi/a)}{j\omega\mu} A \cos\left(\frac{\pi x}{a}\right) \exp(\gamma z)$$

From the other Maxwell equation, $\nabla \times \mathbf{H} = j\omega\epsilon\mathbf{E}$, we can easily obtain

$$\gamma^2 = \left(\frac{\pi}{a}\right)^2 - \omega^2\mu\epsilon$$

Now, in the dielectric filled region the wave is propagating and hence

$$\omega^2\mu\epsilon > \left(\frac{\pi}{a}\right)^2$$

We can therefore write

$$\gamma = \pm j\beta = \pm j\sqrt{\left\{\omega^2\mu\epsilon - \left(\frac{\pi}{a}\right)^2\right\}}$$

giving the field components for this region as

$$E_y = A \sin\left(\frac{\pi x}{a}\right) \exp(\mp j\beta z)$$

$$H_x = \mp \frac{\beta}{\omega\mu} A \sin\left(\frac{\pi x}{a}\right) \exp(\mp j\beta z)$$

$$H_z = -\frac{(\pi/a)}{j\omega\mu} A \cos\left(\frac{\pi x}{a}\right) \exp(\mp j\beta z)$$

where the upper sign denotes a wave propagating in the direction of positive z, that is, in the direction of the incident wave, and the lower sign denotes a wave in the opposite direction.

Now, in the air filled region $\epsilon = \epsilon_0$ and $\omega^2\mu\epsilon_0 < (\pi/a)^2$. We can therefore put

$$\gamma = \pm \alpha = \pm \sqrt{\left\{\left(\frac{\pi}{a}\right)^2 - \omega^2\mu\epsilon_0\right\}}$$

giving the fields

$$E_y = A \sin\left(\frac{\pi x}{a}\right) \exp(\mp \alpha z)$$

$$H_x = \mp \frac{\alpha}{j\omega\mu} A \sin\left(\frac{\pi x}{a}\right) \exp(\mp \alpha z)$$

$$H_z = -\frac{(\pi/a)}{j\omega\mu} A \cos\left(\frac{\pi x}{a}\right) \exp(\mp \alpha z)$$

Note that again we have two signs in the exponent, the upper sign denoting a wave decaying in the direction of positive z and the lower sign denoting a wave decaying in the opposite direction.

Let us now look at the physical situation with reference to *Figure 7.1*. A wave is incident from the left. It hits the left-hand dielectric/air interface and a wave (evanescent) is set up in the air region which will decay to the right. A wave (propagating) is also reflected towards negative z, i.e. to the left. At the right-hand dielectric/air interface the evanescent wave is not zero since the distance s is finite. Hence this field will excite a wave (propagating) in the right-hand dielectric section which will propagate to the right. This evanescent wave will also be 'reflected' resulting in a field decaying to the left in the air-filled region.

Thus we must assume the existence of *two* travelling waves in the left-hand dielectric region, *two* evanescent waves in the air-filled region and *one* wave, travelling to the right in the right-hand dielectric region. We have the fields of these waves; all that remains is to match the boundary conditions at the two dielectric surfaces. Referring to *Figure 7.6*, let us assign the two

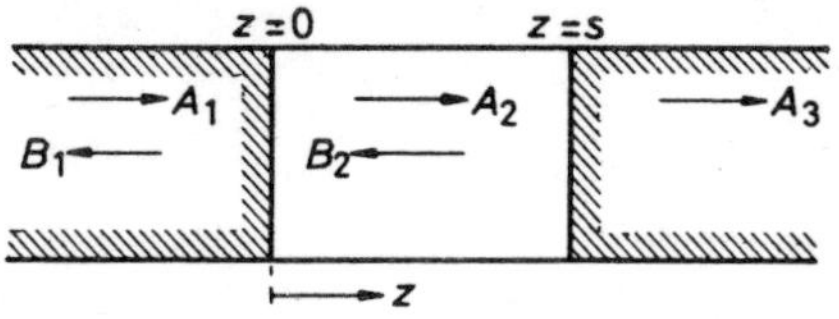

Figure 7.6.

waves at the left amplitudes A_1 and B_1, the two in the air-filled region amplitudes A_2 and B_2 and the wave at the right an amplitude A_3.

Since we are asked to compare powers in the propagating regions and both are effectively the same waveguides then we know that the incident power is proportional to $|A_1|^2$ and the power absorbed in the load is proportional to $|A_3|^2$. We shall therefore calculate the ratio $|A_1/A_3|^2$.

Looking now at the right-hand interface ($z = s$) the boundary conditions require that the total E_y and the total H_x are continuous across the surface. Since the quantity $\sin(\pi x/a)$ is common to all these fields we can ignore this as it will in any case cancel out.

Thus,

$$A_2 \exp(-\alpha s) + B_2 \exp(+\alpha s) = A_3 \exp(-j\beta s)$$

$$-\frac{\alpha}{j\omega\mu}\{A_2 \exp(-\alpha s) - B_2 \exp(+\alpha s)\} = -\frac{\beta}{\omega\mu} A_3 \exp(-j\beta s)$$

which can be solved for A_2 and B_2 giving,

$$A_2 = \frac{1}{2}\left(1 + \frac{j\beta}{\alpha}\right) \exp\{(\alpha - j\beta)s\} A_3$$

$$B_2 = \frac{1}{2}\left(1 - \frac{j\beta}{\alpha}\right) \exp\{-(\alpha + j\beta)s\} A_3$$

The same procedure at the left-hand surface ($z = 0$) gives

$$A_1 + B_1 = A_2 + B_2$$

$$-\frac{\beta}{\omega\mu}(A_1 - B_1) = -\frac{\alpha}{j\omega\mu}(A_2 - B_2)$$

If these are solved for A_1 and the values for A_2 and B_2 in terms of A_3 inserted we readily find that

$$\left|\frac{A_1}{A_3}\right|^2 = \cosh^2 \alpha s + \left\{\frac{\alpha^2 - \beta^2}{2\alpha\beta}\right\}^2 \sinh^2 \alpha s$$

$$= 1 + \left\{\frac{\alpha^2 + \beta^2}{2\alpha\beta}\right\}^2 \sinh^2 \alpha s$$

This is the final answer but the student should note that, if αs is large we can replace $\sinh \alpha s$ by its approximate form ½ exp (αs) when the law becomes

$$\left|\frac{A_1}{A_3}\right|^2 \approx \left\{\frac{\alpha^2 + \beta^2}{4\alpha\beta}\right\}^2 \exp(2\alpha s)$$

The attenuation of such a section of cut-off waveguide becomes linear in decibels with the distance s. Actual instruments have been made on the basis although a circular waveguide is usually used for ease of mechanical construction.

The student, after doing Problem 7.8, may care to try to work out this present problem by transmission line theory, using the 'characteristic impedance' idea of Problem 7.8. For the cut-off section it will be imaginary but, if the analysis is correct, the answer will be the same as that obtained here.

7.8. We must first of all calculate the power, the longitudinal current and the transverse voltage.

(*a*) *Power–P*–This is a well-defined calculation. We integrate the real part of the z-component of the Poynting vector over the cross-section of the guide. Thus

$$\tfrac{1}{2}\,\mathrm{Re}\{\mathbf{k} \cdot \mathbf{E} \times \mathbf{H}^*\} = \frac{1}{2}\frac{\beta}{\omega\mu}|A|^2 \sin^2\left(\frac{\pi x}{a}\right)$$

$$P = \frac{\beta}{2\omega\mu} |A|^2 \int_0^b dy \int_0^a \sin^2\left(\frac{\pi x}{a}\right) dx = \frac{ab\beta}{4\omega\mu} |A|^2$$

(*b*) *Longitudinal current–I*–Longitudinal current flows only on the broad faces of the waveguide for this mode, the surface current density in the *z*-direction being equal in magnitude to H_x, positive or negative according to which broad face is under consideration. Thus

$$I = \int_{x=0}^{a} J_z \, dx = \frac{2a\beta}{\pi\omega\mu} A$$

(*c*) *Transverse voltage–V*–This is a rather arbitrary choice. As stated in the problem we choose (and this is the conventional choice) the line integral of the electric field from one broad face to the other at the *centre* of the waveguide. Since the electric field is independent of y the calculation is trivial and

$$V = bA$$

If we now define the three impedances as in the problem we have

$$Z_{VI} = \frac{V}{I}$$

$$Z_{PI} = \frac{2P}{|I|^2}$$

$$Z_{PV} = \frac{|V|^2}{2P}$$

Substituting the values we have determined we obtain

$$Z_{PI} = \frac{\pi^2}{8}\frac{\omega\mu}{\beta}\cdot\frac{b}{a} = \frac{\pi^2}{8}\frac{k}{\beta}\cdot\frac{b}{a}\cdot\sqrt{\frac{\mu}{\epsilon}}$$

$$Z_{PV} = 2\frac{\omega\mu}{\beta}\cdot\frac{b}{a} = 2\frac{k}{\beta}\cdot\frac{b}{a}\cdot\sqrt{\frac{\mu}{\epsilon}}$$

$$Z_{VI} = \frac{\pi}{2}\frac{\omega\mu}{\beta}\cdot\frac{b}{a} = \frac{\pi}{2}\cdot\frac{k}{\beta}\cdot\frac{b}{a}\cdot\sqrt{\frac{\mu}{\epsilon}}$$

We can observe that these are all different values to the extent of a numerical constant. The dependence on the various physical parameters, however, is the same and useful conclusions can be drawn from this property.

Now the curious student may question the unsatisfactory definition of transverse voltage which is, in fact, a very arbitrary choice. Is there any possibility that we have chosen wrongly and that this is why the answers are inconsistent? Again the curious student may come to the conclusion that if the voltage had turned out to be

$$\frac{\pi}{4}bA$$

then all the impedances would take the value of Z_{PI} and would then be identical. Can we justify the $\pi/4$? The mean value of $\sin(\pi x/a)$ is $2/\pi$ so that it is not just the mean voltage which is required. Fortunately the question can be resolved in a somewhat indirect way. If this discussion has any meaning then it should be possible to carry out the same sort of discussion for *any* mode in a waveguide. Consider, however, the TE_{01} mode in a circular guide (we have calculated these fields in Problem 7.4). The flow lines of the electric field are circular and it is therefore not possible to carry out an integration from one wall to another. Transverse voltage in this case has no meaning at all. Worse, the wall currents are entirely circumferential and hence I is zero.

It would therefore appear rather pointless to try to find a unique value of characteristic impedance for a waveguide. Notwithstanding this discussion the *functional* dependence for the TE_{10} rectangular waveguide mode has found useful applications as typified in the next problem.

7.9. Based on the calculations of the last problem all impedances were shown to be proportional to b and we would therefore expect

$$\frac{Y_0'}{Y_0} = \frac{b}{b'}$$

and this is found to be so.

The next part of the problem will be recognised from transmission line theory as the standard quarter wave transformer in which

$$s = \lambda/4$$

$$Y_{02}^2 = Y_{01}.Y_{03}$$

are sufficient to satisfy the matching condition. Using the result of the first part yields for b_2

$$b_2 = \sqrt{(b_1 b_3)}$$

7.10. This structure is a form of resonator used in slab or stripline systems. The upper and lower halves are identical and we need to consider only one of them. If d is small the electric field will have only one component, directly from one plate to the other and if we are to neglect fringing, the edges of the plates are bounded by magnetic walls.

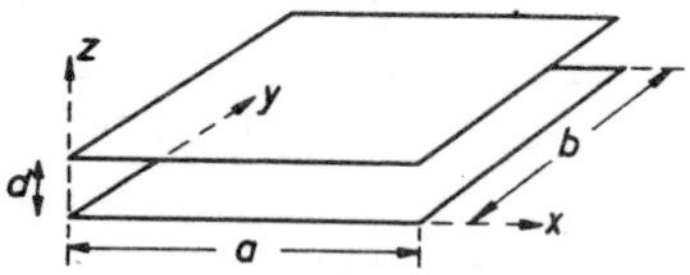

Figure 7.7.

We are therefore considering a rectangular box as shown in *Figure 7.7* with conducting walls at $z = 0$ and $z = d$ and with magnetic walls at $x = 0$, $x = a$, $y = 0$ and $y = b$. The electric field has the one component E_z, the magnetic field having components H_x and H_y.

Since $\nabla . \mathbf{D} = 0$, then $\nabla . \mathbf{E} = 0$ and we find that E_z is independent of z. From $\nabla \times \mathbf{H} = j\omega\epsilon\mathbf{E}$ we can likewise show that both H_x and H_y are independent of z. All the field components are therefore functions of x and y alone.

Maxwell's equations reduce to

$$\frac{\partial E_z}{\partial y} = -\mathrm{j}\omega\mu H_x$$

$$\frac{\partial E_z}{\partial x} = \mathrm{j}\omega\mu H_y$$

$$\frac{\partial H_y}{\partial x} - \frac{\partial H_x}{\partial y} = \mathrm{j}\omega\epsilon E_z$$

and substituting for H_x and H_y in the third we find that E_z satisfies the differential equation

$$\frac{\partial^2 E_z}{\partial x^2} + \frac{\partial^2 E_z}{\partial y^2} + \omega^2\mu\epsilon E_z = 0$$

This is first of all solved by the technique of separation of variables, a standard procedure leading to

$$E_z = X(x)Y(y)$$

with

$$X(x) = A \sin(k_x x) + B \cos(k_x x)$$

$$Y(y) = C \sin(k_y y) + D \cos(k_y y)$$

the parameters k_x and k_y being related by

$$k_x^2 + k_y^2 = \omega^2\mu\epsilon$$

We can now apply the boundary conditions. On the magnetic walls $x = 0$ and $x = a$, the magnetic field, $H_y = 0$. From the second of Maxwell's equations above this means that

$$\frac{\mathrm{d}X}{\mathrm{d}x} = 0 \qquad \text{for } x = 0 \text{ and } x = a$$

Hence $$A = 0 \quad \text{and} \quad k_x a = m\pi \qquad (m \text{ being an integer})$$

Similarly on $y = 0$ and $y = b$, $H_x = 0$. This leads to

$$C = 0 \quad \text{and} \quad k_y b = n\pi \qquad (n \text{ being an integer})$$

Collecting this information together and substituting the values of k_x and k_y into the equation relating them gives the resonant frequencies of the plate as

$$\omega^2 \mu \epsilon = \left(\frac{m\pi}{a}\right)^2 + \left(\frac{n\pi}{b}\right)^2$$

CHAPTER 8

8.1. The first part of this problem is bookwork and we restrict ourselves to the second part for which the geometry is as illustrated in *Figure 8.2.*

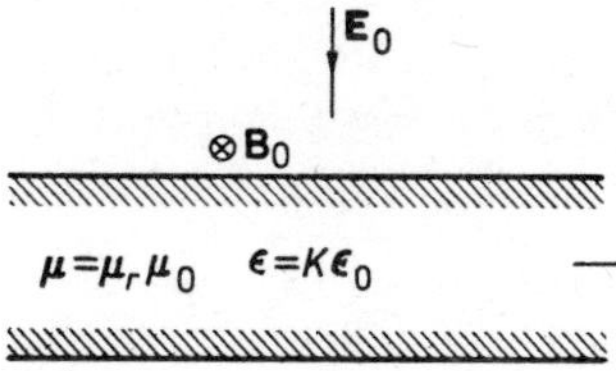

Figure 8.2.

Let us attach a moving coordinate system Σ' to the plate. An observer stationary in this system will observe an electric and a magnetic field (due to $\mathbf{E}_0$ and $\mathbf{B}_0$) and will measure them as

$$\mathbf{E}_0' = \mathbf{E}_0 + \mathbf{v} \times \mathbf{B}_0$$

$$\mathbf{B}_0' = \mathbf{B}_0 - \frac{\mathbf{v} \times \mathbf{E}_0}{c^2}$$

These are effectively the fields impressed on the space from the external sources and are applied to a now stationary dielectric. Thus $\mathbf{D}'$ and $\mathbf{H}'$ must be continuous across the surface of the dielectric since $\mathbf{E}_0'$ is normal to and $\mathbf{B}_0'$ is tangential to the surface. Hence, if $\mathbf{E}'$ and $\mathbf{B}'$ represent the fields *inside* the dielectric we must satisfy

$$\epsilon_0 \mathbf{E}_0' = \epsilon \mathbf{E}'$$

$$\frac{1}{\mu_0}\mathbf{B}_0' = \frac{1}{\mu}\mathbf{B}'$$

or

$$\mathbf{E}' = \frac{\epsilon_0}{\epsilon}\mathbf{E}_0'$$

$$\mathbf{B}' = \frac{\mu}{\mu_0}\mathbf{B}_0'$$

Now these fields inside the dielectric can be considered to be composed of two parts which must be separated before we transform back to the original system. Of $\mathbf{E}'$, $\mathbf{E}_0'$ is the original field of the external source and $\mathbf{E}_d' = (\mathbf{E}' - \mathbf{E}_0')$ is an 'excess' field due to the polarization of the dielectric. Similarly, of $\mathbf{B}'$, $\mathbf{B}_0'$ is the original field and $\mathbf{B}_d' = (\mathbf{B}' - \mathbf{B}_0')$ is an 'excess' field due to the magnetization of the dielectric. Now of these two parts of each field, the parts $\mathbf{E}_0'$ and $\mathbf{B}_0'$ will, if the equations were consistent, transform back to the original $\mathbf{E}_0$ and $\mathbf{B}_0$. It is the 'excess' fields which we must study to determine the effect of the moving dielectric. Consequently let us see how an observer in the stationary system Σ would measure the fields $\mathbf{E}_d'$ and $\mathbf{B}_d'$ which only exist in the dielectric filled region. Since the system Σ is moving relative to Σ' with a velocity $-\mathbf{v}$ we can write immediately that, in the dielectric region an observer in Σ would measure fields $\mathbf{E}_d$ and $\mathbf{B}_d$ given by

$$\mathbf{E}_d = \mathbf{E}_d' - \mathbf{v} \times \mathbf{B}_d'$$

$$\mathbf{B}_d = \mathbf{B}_d' + \frac{\mathbf{v} \times \mathbf{E}_d'}{c^2}$$

The question only asks for the polarization $\mathbf{P}$ so that we can ignore $\mathbf{B}_d$.

Now the total electric field in the dielectric is just $\mathbf{E}_0 + \mathbf{E}_d = \mathbf{E}$ (say). Further, in the dielectric $\mathbf{D}$ must be equal to $\epsilon_0\mathbf{E}_0$, and the observer in Σ would therefore say that the dielectric had a polarization $\mathbf{P}$ given by

$$\mathbf{P} = \epsilon_0(\mathbf{E}_0 - \mathbf{E}) = -\epsilon_0\mathbf{E}_d$$

If we now substitute for $\mathbf{E}_d$ and let $\epsilon = K\epsilon_0$ and $\mu = \mu_r\mu_0$ (note the slight deviation from the notation for the permeability used in the question) we obtain

$$\mathbf{P} = \epsilon_0\left\{\left(1 - \frac{1}{K}\right)\mathbf{E}_0' + (\mu_r - 1)\,\mathbf{v} \times \mathbf{B}_0'\right\}$$

and, making a final substitution for $\mathbf{E}_0'$ and $\mathbf{B}_0'$ in terms of $\mathbf{E}_0$ and $\mathbf{B}_0$, we find

$$\mathbf{P} = \epsilon_0\left\{\left(1 - \frac{1}{K}\right)(\mathbf{E}_0 + \mathbf{v} \times \mathbf{B}_0) + (\mu_r - 1)\,\mathbf{v} \times \left(\mathbf{B}_0 - \frac{\mathbf{v} \times \mathbf{E}_0}{c^2}\right)\right\}$$

To be consistent with our non-relativistic equations, the question gives $v \ll c$, so we must neglect terms of order $(v/c)^2$ which means that the very last terms in the bracket should be neglected, whence, as required,

$$\mathbf{P} = \epsilon_0\left\{\left(1 - \frac{1}{K}\right)\mathbf{E}_0 + \left(\mu_r - \frac{1}{K}\right)\mathbf{v} \times \mathbf{B}_0\right\}$$

8.2. This problem is a combination of experiments carried out by Röntgen and Eichenwald about the year 1900. If carefully studied they lead to a very good understanding of the behaviour of moving systems and the problem is included for this purpose.

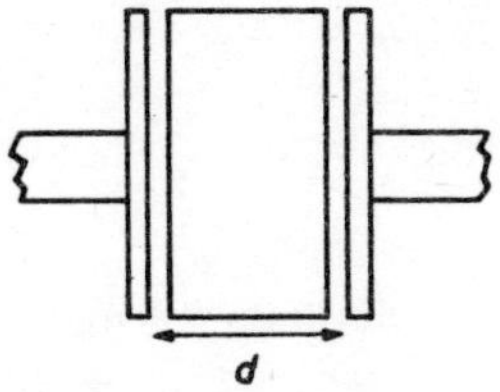

Figure 8.3.

Consider the geometry shown in *Figure 8.3.* The key to the solution is to regard the electric field **E**, equal in magnitude to V/d, as being composed of two parts, one signified by $\mathbf{E}_p$ due to the charge on the plates and the other $\mathbf{E}_d$ due to the polarization of the dielectric. In order to make this clear suppose that there are air gaps between the dielectric and the plates as shown in the figure. Suppose also that the gaps are negligible in relation to d so that $|\mathbf{E}| = V/d$ is not affected. Then $\mathbf{E}_p$ is the field in the air gap and **E** the total field in the dielectric composed of $\mathbf{E}_p$ and $\mathbf{E}_d$. Thus

$$\mathbf{E} = \mathbf{E}_p + \mathbf{E}_d$$

Since $$\mathbf{D}_p = \epsilon_0 \mathbf{E}_p = \mathbf{D} = \epsilon \mathbf{E},$$

then $$\mathbf{E}_p = \frac{\epsilon}{\epsilon_0}\mathbf{E} \quad \text{and} \quad \mathbf{E}_d = -\left(\frac{\epsilon}{\epsilon_0} - 1\right)\mathbf{E}$$

Note also that, since $\mathbf{D}_p = \mathbf{D}$, $\mathbf{D}_d = 0$ and no flux density vector results from $\mathbf{E}_d$.

After these preliminary comments we are in a position to consider the three cases of the problem.

(*a*) If the plates only move then the moving fields are $\mathbf{E}_p$ and $\mathbf{D}_p$. It is these fields which will produce any magnetic effects measured by a stationary observer. Now suppose that the edges of the plates move with a velocity **v**. Then the stationary observer can be regarded as moving with a velocity $-\mathbf{v}$

relative to the fields $\mathbf{E}_p$ and $\mathbf{D}_p$. He will therefore measure magnetic fields (using the inverse equations of the introduction) given by

$$\mathbf{B} = \frac{\mathbf{v} \times \mathbf{E}_p}{c^2} = \left(\frac{\epsilon}{\epsilon_0}\right) \cdot \frac{\mathbf{v} \times \mathbf{E}}{c^2} = \mu_0 \epsilon \mathbf{v} \times \mathbf{E}$$

$$\mathbf{H} = \mathbf{v} \times \mathbf{D}_p = \mathbf{v} \times \mathbf{D} = \epsilon \mathbf{v} \times \mathbf{E}$$

[using $c^2 = (\mu_0 \epsilon_0)^{-1}$]

Let us leave these equations until later.

(*b*) If the dielectric only moves then the moving fields are $\mathbf{E}_d$ and $\mathbf{D}_d$, the latter being zero as already noted. The magnetic effects measured by the stationary observer are then

$$\mathbf{B} = \frac{\mathbf{v} \times \mathbf{E}_d}{c^2} = -\left(\frac{\epsilon}{\epsilon_0} - 1\right) \frac{\mathbf{v} \times \mathbf{E}}{c^2} = -\mu_0 (\epsilon - \epsilon_0)\, \mathbf{v} \times \mathbf{E}$$

$$\mathbf{H} = 0$$

(*c*) Finally, if both the plates and the dielectric move then the moving fields are **E** and **D**, with magnetic effects

$$\mathbf{B} = \frac{\mathbf{v} \times \mathbf{E}}{c^2} = \mu_0 \epsilon_0\, \mathbf{v} \times \mathbf{E}$$

$$\mathbf{H} = \mathbf{v} \times \mathbf{D} = \epsilon \mathbf{v} \times \mathbf{E}$$

We have now found the three formal sets of solutions. Note that in case (*b*) the magnetic field $\mathbf{H} = 0$ and any **B** would be interpreted as due to a magnetization, **M**. (Note also that **B** is in the opposite direction to that of the other two cases.)

Now the observer in the stationary system would write

$$\mathbf{B} = \mu_0 (\mathbf{H} + \mathbf{M})$$

and we can calculate **M** for the three cases since we already have **B** and **H**. If we do this and tabulate the results we can correlate the apparent effects with the moving materials. The table is:

Case	*Moving objects*	**H**	**M**
(*a*)	Plates only	$\epsilon \mathbf{v} \times \mathbf{E}$	0
(*b*)	Dielectric only	0	$-(\epsilon - \epsilon_0)\mathbf{v} \times \mathbf{E}$
(*c*)	Both	$\epsilon \mathbf{v} \times \mathbf{E}$	$-(\epsilon - \epsilon_0)\mathbf{v} \times \mathbf{E}$

Expressed this way we can see that if the plates move, whether alone or not, they can be regarded as producing a magnetic field $\mathbf{H} = \epsilon \mathbf{v} \times \mathbf{E}$ and when the dielectric moves, whether alone or not, it can be regarded as producing a magnetization $\mathbf{M} = -(\epsilon - \epsilon_0)\mathbf{v} \times \mathbf{E}$. Note that the charges on the plates are free charges and hence a motion of these would produce a current in the free sense which would be considered to produce a magnetic field and not magnetization. The charges associated with the dielectric, however, are bound charges (polarization charge) and their motion produces a current more akin to the atomic currents responsible for magnetization.

8.3. Let us attach a coordinate system Σ' to the moving liquid and a system Σ to the stationary parts. The magnetic field $\mathbf{B}$ in Σ would be observed in Σ' to be an electric field $\mathbf{E}'$ and a magnetic field $\mathbf{B}'$ given by

$$\mathbf{E}' = \mathbf{v} \times \mathbf{B}$$

$$\mathbf{B}' = \mathbf{B}$$

Now the electric field $\mathbf{E}'$ is applied to the whole of the Σ' space and will therefore polarize the dielectric, and the electric field in the dielectric, denoted by $\mathbf{E}'_d$ can be calculated from the continuity of $\mathbf{D}'$ as

$$\mathbf{E}'_d = \frac{\epsilon_0}{\epsilon}\mathbf{E}' = \frac{1}{\epsilon_r}\mathbf{E}'$$

This field $\mathbf{E}'_d$ can be regarded as composed of the field $\mathbf{E}'$ plus an 'excess' field $\mathbf{E}'_e$ and an observer in the stationary system Σ will notice an electric field due to this excess field and none due to the field $\mathbf{E}'$. The electric field in the stationary system is

$$\mathbf{E}_e = \mathbf{E}'_e$$

and this field will appear across the electrodes. Thus, if t is the dimension across the tube then the voltage V developed is

$$V = tE_e = t\left(1 - \frac{1}{\epsilon_r}\right) vB$$

Inserting the numerical values given shows that

$$V = 9{\cdot}6 \text{ mV}$$

8.4. The electric field in the stationary system can be written as

$$|E| = \frac{V \sin \omega\tau}{t}$$

Now this is precisely the same case as the moving dielectric (case (*b*)) of Problem 8.2 and hence the magnetic field **B** measured in the stationary system (which you will remember is similar to magnetization) is just

$$\mathbf{B} = -\left(\frac{\epsilon}{\epsilon_0} - 1\right)\frac{\mathbf{v} \times \mathbf{E}}{c^2}$$

or, in magnitude,

$$|B| = \frac{(\epsilon_r - 1)\, vV \sin \omega\tau}{tc^2}$$

Thus the total flux Φ linked with the coil is

$$\Phi = NAB = \frac{NA(\epsilon_r - 1)vV}{tc^2} \sin \omega\tau$$

and the electromotive force induced is given by

$$\text{e.m.f.} = -\frac{d\Phi}{d\tau} = -\frac{\omega NA(\epsilon_r - 1)vV}{tc^2} \cos \omega\tau$$

Inserting the numerical values gives the final answer

$$\text{e.m.f.} = 0{\cdot}044 \text{ mV peak}$$

8.5. The first part of this problem is standard bookwork and we shall proceed to the second part.

This is in three sections, each section being an application of the expression of Faraday's Law as given in the question, since

$$V = \oint_{\text{loop}} \mathbf{E} \,.\, d\mathbf{l} = \int_s \nabla \times \mathbf{E} \,.\, d\mathbf{s}$$

The value of the field **B** can be determined from I by elementary means. If ρ measures the radial distance of an arbitrary point from the wire, then

$$\mathbf{B} = \mathbf{a}_\phi \left(\frac{\mu_0 I}{2\pi\rho}\right)$$

Now consider the three sections.

(*i*) in this case $\mathbf{u} = 0$ and **B** is a function of time. Thus

$$\mathbf{B} = \mathbf{a}_\phi \frac{\mu_0 I_0}{2\pi\rho} \cos \omega t$$

and

$$\nabla \times \mathbf{E} = -\frac{\partial \mathbf{B}}{\partial t} = \mathbf{a}_\phi \frac{\omega\mu_0 I_0}{2\pi\rho} \sin \omega t$$

For the loop,

$$\mathrm{d}\mathbf{s} = \mathbf{a}_\phi \, \mathrm{d}\rho \, \mathrm{d}z$$

$$\therefore \qquad \nabla \times \mathbf{E} \, . \, \mathrm{d}\mathbf{s} = \frac{\omega\mu_0 I_0}{2\pi\rho} \sin \omega t \, \mathrm{d}\rho \, \mathrm{d}z$$

leading to

$$V = \frac{\omega\mu_0 I_0}{2\pi} \sin \omega t \int_{z=0}^{a} \mathrm{d}z \int_{\rho=r}^{r+b} \frac{\mathrm{d}\rho}{\rho} = \frac{\omega\mu_0 I_0 a}{2\pi} \ln \left\{1 + \frac{b}{r}\right\} \sin \omega t$$

(*ii*) In this case **B** is independent of time but, say,

$$\mathbf{u} = -u_0 \mathbf{a}_\rho$$

Thus

$$\mathbf{u} \times \mathbf{B} = -\mathbf{a}_z \frac{u_0 \mu_0 I_0}{2\pi\rho}$$

and

$$\nabla \times \mathbf{E} = \nabla \times (\mathbf{u} \times \mathbf{B}) = -\mathbf{a}_\phi \frac{u_0 \mu_0 I_0}{2\pi\rho^2}$$

Therefore,

$$\nabla \times \mathbf{E} \,.\, \mathrm{ds} = -\frac{u_0 \mu_0 I_0}{2\pi\rho^2}\, \mathrm{d}\rho\, \mathrm{d}z$$

leading to

$$V = -\frac{u_0 \mu_0 I_0}{2\pi} \int_0^a \mathrm{d}z \int_r^{b+r} \frac{\mathrm{d}\rho}{\rho^2} = -\frac{u_0 \mu_0 I_0 a}{2\pi} \left\{ \frac{1}{r} - \frac{1}{r+b} \right\}$$

(*iii*) In this case there is both motion and time variation. Since

$$\mathbf{B} = \mathbf{a}_\phi \frac{\mu_0 I_0}{2\pi\rho} \cos \omega t$$

we obtain

$$\frac{\partial \mathbf{B}}{\partial t} = -\mathbf{a}_\phi \frac{\omega \mu_0 I_0}{2\pi\rho} \sin \omega t$$

$$\nabla \times (\mathbf{u} \times \mathbf{B}) = -\mathbf{a}_\phi \frac{u_0 \mu_0 I_0}{2\pi\rho^2} \cos \omega t$$

whence

$$\nabla \times \mathbf{E} \,.\, \mathrm{ds} = \frac{\mu_0 I_0}{2\pi} \left\{ \frac{\omega}{\rho} \sin \omega t - \frac{u_0}{\rho^2} \cos \omega t \right\} \mathrm{d}z\, \mathrm{d}\rho$$

leading to

$$V = \frac{a \mu_0 I_0}{2\pi} \left\{ \omega \ln \left[1 + \frac{b}{r} \right] \sin \omega t - u_0 \left[\frac{1}{r} - \frac{1}{r+b} \right] \cos \omega t \right\}$$

While this formally answers the question using the information given, it is also of interest to look at the second part with an alternative approach. Due to the field **B**, which is in the stationary system an observer moving with the coil would observe an electric field **E**′ given by

$$\mathbf{E}' = \mathbf{u} \times \mathbf{B} = -\mathbf{a}_z \frac{u_0 \mu_0 I_0}{2\pi\rho}$$

This electric field will cause charge to flow in the two vertical wires only, to the extent that the resultant electric field in these wires is reduced to zero since they are conductors. The field due to these charges can be considered as an excess field which is, in fact, equal and opposite to **E**′. This field, as a glance at the transformation equations will show, will be observed in the stationary system. (The original field $\mathbf{E}'$ will not be observed since it came from the magnetic field **B**.)

Thus the vertical section of wire at r will appear to have an electric field along it equal to

$$-\mathbf{E}' = \mathbf{a}_z \frac{u_0 \mu_0 I_0}{2\pi r}$$

and hence a voltage

$$\frac{u_0 \mu_0 I_0 a}{2\pi r}$$

the lower end being positive.

Likewise the vertical section at $(b + r)$ will have a voltage

$$\frac{u_0 \mu_0 I_0 a}{2\pi(b + r)}$$

the lower end again being positive.

The net voltage is the difference between these two and since in the first treatment of this case the element **dS** was taken in the direction of $\mathbf{a}_\phi$ so that the line integral was taken clockwise around the loop as it appears in *Figure 8.1.* The net voltage in the same direction is

$$V = V(b + r) - (V/r) = \frac{u_0 \mu_0 I_0 a}{2\pi} \left\{ \frac{1}{b + r} - \frac{1}{r} \right\}$$

as before.

8.6. Consider the geometry of *Figure 8.4.* Inside the magnet **B** is in the z-direction of a cylindrical coordinate system. The vector potential **A** has a

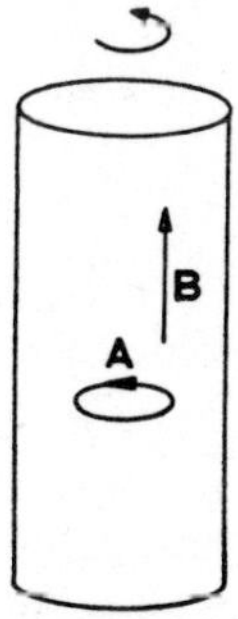

Figure 8.4.

component A_ϕ only. If the magnet now rotates, an observer who is stationary will observe an electric field

$$\mathbf{E} = -\mathbf{v} \times \mathbf{B} = -\mathbf{v} \times (\nabla \times \mathbf{A})$$

which can be expanded by the rules of vector algebra to give

$$\mathbf{E} = -\nabla(\mathbf{v} \cdot \mathbf{A}) + \{\mathbf{A} \times (\nabla \times \mathbf{v}) + (\mathbf{A} \cdot \nabla)\mathbf{v} + (\mathbf{v} \cdot \nabla)\mathbf{A}\}$$

Now, since **v** and **A** are parallel, $\mathbf{v} \times \mathbf{A} = 0$, whence

$$\nabla \times (\mathbf{v} \times \mathbf{A}) = \mathbf{v}\,\nabla \cdot \mathbf{A} - \mathbf{A}\,\nabla \cdot \mathbf{v} + (\mathbf{A} \cdot \nabla)\mathbf{v} - (\mathbf{v} \cdot \nabla)\mathbf{A} = 0$$

Since A is solenoidal, $\nabla \cdot \mathbf{A} = 0$ and likewise $\nabla \cdot \mathbf{v} = 0$. Therefore

$$(\mathbf{A} \cdot \nabla)\mathbf{v} = (\mathbf{v} \cdot \nabla)\mathbf{A}$$

Also since $\mathbf{v} = \boldsymbol{\omega} \times \mathbf{r}$, $\nabla \times \mathbf{v} = 2\boldsymbol{\omega}$ and the term of **E** which is enclosed in brackets becomes

$$2\{\mathbf{A} \times \boldsymbol{\omega} + (\mathbf{A} \cdot \nabla)\mathbf{v}\}$$

It is now readily shown that

$$\mathbf{A} \times \boldsymbol{\omega} = \mathbf{a}_r\, \omega A_\phi$$

and that

$$(\mathbf{A} \cdot \nabla)\mathbf{v} = -\mathbf{a}_r\, \omega A_\phi$$

and this term therefore vanishes, leaving

$$\mathbf{E} = -\nabla(\mathbf{v} \cdot \mathbf{A})$$

8.7. Consider the geometry of *Figure 8.5.*

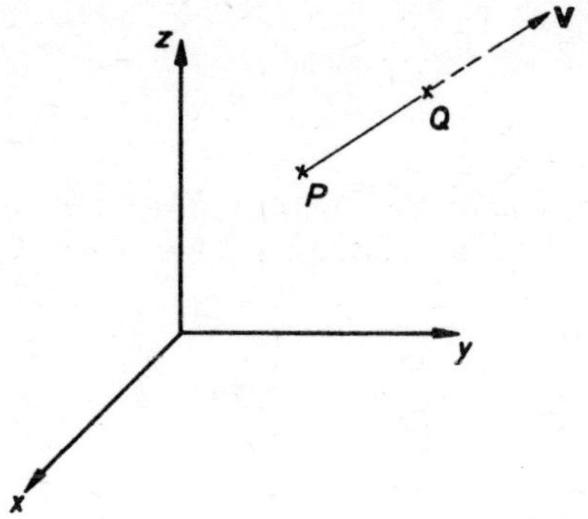

Figure 8.5.

Let Q be a point fixed in the coordinate system and P be a point fixed in the body such that if P has its shown position at time t it will have moved to Q (in the coordinate system) in a time $\mathrm{d}t$. Thus, say,

$$PQ = \mathbf{v}\,\mathrm{d}t = \mathbf{i}\,\mathrm{d}x + \mathbf{j}\,\mathrm{d}y + \mathbf{k}\,\mathrm{d}z$$

Now if the vector field $\mathbf{A}$ has a value $\mathbf{A}$ at Q at time t, then after a time $\mathrm{d}t$ it will have changed to $\mathbf{A} + \mathrm{d}\mathbf{A}$ which, since $\mathbf{A}$ is fixed in the body, will be the same value of the field at P and at time t.

Thus

$$A_x - (A_x + \mathrm{d}A_x) = \left[\frac{\partial A_x}{\partial x}\,\mathrm{d}x + \frac{\partial A_x}{\partial y}\,\mathrm{d}y + \frac{\partial A_x}{\partial z}\,\mathrm{d}z\right]$$

with similar expressions for the y and z-components. The rate of change of A_x with t is therefore

$$\frac{\partial A_x}{\partial t} = -\left[\frac{\partial A_x}{\partial x}\frac{\mathrm{d}x}{\mathrm{d}t} + \frac{\partial A_x}{\partial y}\frac{\mathrm{d}y}{\mathrm{d}t} + \frac{\partial A_x}{\partial z}\frac{\mathrm{d}z}{\mathrm{d}t}\right] = -\mathbf{v}\,.\,\boldsymbol{\nabla} A_x$$

with similar expressions for the other components. Adding the components we obtain

$$\frac{\partial \mathbf{A}}{\partial t} = -\mathbf{v}\,.\,\boldsymbol{\nabla}\,[\mathbf{i}\,A_x + \mathbf{j}\,A_y + \mathbf{k}\,A_z] = -(\mathbf{v}\,.\,\boldsymbol{\nabla})\mathbf{A}$$

8.8. Let us consider the electric field in the stationary system. In terms of vector and scalar potentials this can be written as

$$\mathbf{E} = -\frac{\partial \mathbf{A}}{\partial t} - \nabla \phi$$

Now, if **A** is the vector potential of the current, then in terms of Problem 8.7 it is fixed in the moving body, and by the solution of this previous problem

$$\frac{\partial \mathbf{A}}{\partial t} = -(\mathbf{v} \cdot \nabla)\mathbf{A}$$

whence,

$$\mathbf{E} = (\mathbf{v} \cdot \nabla)\mathbf{A} - \nabla \phi$$

But we know that the motion produces an electric field given by

$$\mathbf{E} = -\mathbf{v} \times \mathbf{B} = -\mathbf{v} \times (\nabla \times \mathbf{A})$$

Now, from vector algebra,

$$\nabla(\mathbf{v} \cdot \mathbf{A}) = (\mathbf{v} \cdot \nabla)\mathbf{A} + (\mathbf{A} \cdot \nabla)\mathbf{v} + \mathbf{v} \times (\nabla \times \mathbf{A}) + \mathbf{A} \times (\nabla \times \mathbf{v})$$

and if **v** is a *constant* velocity vector

$$\nabla(\mathbf{v} \cdot \mathbf{A}) = (\mathbf{v} \cdot \nabla)\mathbf{A} + \mathbf{v} \times (\nabla \times \mathbf{A})$$

whence, substituting for $\mathbf{v} \times (\nabla \times \mathbf{A})$

$$\mathbf{E} = (\mathbf{v} \cdot \nabla)\mathbf{A} - \nabla(\mathbf{v} \cdot \mathbf{A})$$

Comparing this with the potential expression for **E** we see that

$$\nabla \phi = \nabla(\mathbf{v} \cdot \mathbf{A})$$

or

$$\phi = \mathbf{v} \cdot \mathbf{A} + \text{const.}$$

Since, when **v** is zero, ϕ is zero, then the constant is zero and

$$\phi = \mathbf{v} \cdot \mathbf{A}$$

Now let us calculate the vector potential of a small element $\mathbf{dl} = \mathbf{a}_t \, dl$ of the loop. At a point a distance r from the element,

$$\mathbf{A} = \frac{\mu_0 I}{4\pi r} \mathbf{dl} = \frac{\mu_0 \mathbf{I}}{4\pi r} dl$$

where **I** signifies $I\mathbf{a}_t$

Thus the potential ϕ is given by

$$\phi = \frac{\mu_0 \mathbf{v} \cdot \mathbf{I}}{4\pi r} dl$$

Now if the loop is considered to carry an apparent charge density σ then the element dl will have a charge $\sigma \, dl$ and the potential due to this charge at a distance r will be

$$\phi = \frac{\sigma}{4\pi\epsilon_0 r} dl$$

Since these two expressions for potential should be the same then

$$\mu_0 \mathbf{v} \cdot \mathbf{I} = \frac{\sigma}{\epsilon_0}$$

or

$$\sigma = \frac{\mathbf{v} \cdot \mathbf{I}}{c^2}$$

as required.

To solve the second part of the problem consider *Figure 8.6.* The fact that $b \ll a$ means that fringing effects are to be neglected.

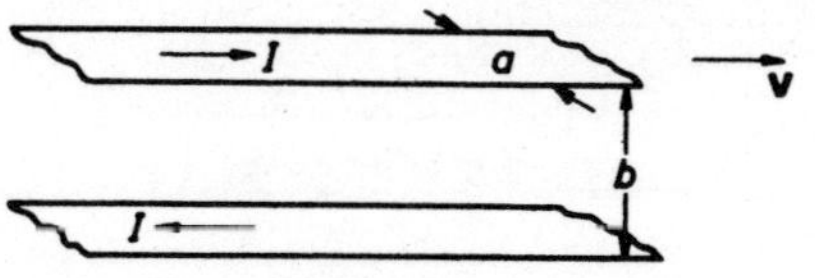

Figure 8.6.

Since **v** is in the same direction as **I**, the apparent charge per unit length is just

$$\sigma = \frac{vI}{c^2}$$

The surface charge density is therefore

$$\frac{\sigma}{a} = \frac{vI}{ac^2}$$

and the electric field is

$$E = \frac{vI}{\epsilon_0 ac^2} = \frac{\mu_0 vI}{a}$$

obtained from this concept of apparent charge.

Now, in a system of coordinates moving with the transmission line an observer would detect only a magnetic field of strength

$$B' = \frac{\mu_0 I}{a}$$

and this would be in a direction transverse to the strip. Now a stationary observer can be considered as moving with a velocity $-\mathbf{v}$ relative to the line and hence will observe an electric field

$$\mathbf{E} = -\mathbf{v} \times \mathbf{B}'$$

which is perpendicular to **v** and to $\mathbf{B}'$ and is therefore across the b-dimension of the strip. Since **v** and $\mathbf{B}'$ are perpendicular this electric field is just equal to vB' in magnitude and hence

$$|E| = \frac{\mu_0 I}{a} v \quad \text{as before}$$

The student should check that the directions of both electric fields are the same.

REFERENCES

The following list of books should be considered as background reading, especially for particular topics.

1. J. A. Stratton, *Electromagnetic Theory,* McGraw-Hill, 1941
2. S. Ramo and J. R. Whinnery, *Fields and Waves in Modern Radio,* Wiley, 1953
3. Feynmann, *Lectures on Physics,* Vol. 2, Addison-Wesley, 1964. Particularly for elementary approach to variational techniques
4. E. G. Cullwick, *Electromagnetism and Relativity,* Longmans, 1957. For moving media
5. R. M. Fano, L. J. Chu and R. B. Adler, *Electromagnetic Fields, Energy and Forces,* Wiley, 1960
6. R. B. Adler, L. J. Chu and R. M. Fano, *Electromagnetic Energy Transmission and Radiation,* Wiley, 1960. With parts of Reference 5 this gives a good coverage of circuit concepts
7. R. E. Collin, *Foundations for Microwave Engineering,* McGraw-Hill, 1966. An excellent book on waveguide technique with a good selection of problems
8. R. E. Collin, *Field Theory of Guided Waves,* McGraw-Hill, 1960. An advanced book but some sections are well worth reading
9. M. R. Spiegel, *Theory and Problems of Vector Algebra,* Schaum's Outline Series, Schaum, 1959. A full coverage of vector algebra required for field theory
10. H. Margenau and G. M. Murphy, *The Mathematics of Physics and Chemistry,* Van Nostrand, 1943. Covers the special functions required